RSAC

From Bauhaus to Ecohouse

FROM BAUHAUS TO ECOHOUSE

A HISTORY OF ECOLOGICAL DESIGN

PEDER ANKER

LOUISIANA STATE UNIVERSITY PRESS
BATON ROUGE

PUBLISHED BY LOUISIANA STATE UNIVERSITY PRESS
Copyright © 2010 by Louisiana State University Press
All rights reserved
Manufactured in the United States of America
First printing

DESIGNER: Michelle A. Neustrom
TYPEFACES: Chapparral Pro, Trade Gothic, Museo
PRINTER AND BINDER: Thomson-Shore, Inc.

Library of Congress Cataloging-in-Publication Data
Anker, Peder.
    From Bauhaus to ecohouse : a history of ecological design / Peder Anker.
        p. cm.
    Includes bibliographical references and index.
    ISBN 978-0-8071-3551-8 (cloth : alk. paper)
    1. Architecture—Environmental aspects. 2. Architecture and science—History—20th century. 3. Architecture, Modern—20th century.
    I. Title.
    NA2542.35.A54 2010
    720'.47—dc22
                                                    2009020217

The paper in this book meets the guidelines for permanence and durability of the Committee on Production Guidelines for Book Longevity of the Council on Library Resources. ∞

CONTENTS

ACKNOWLEDGMENTS > vii

Introduction > 1

1 The Bauhaus of Nature > 9
2 Planning the Economy of Nature > 24
3 The New American Bauhaus of Nature > 37
4 The Graphic Environment of Herbert Bayer > 54
5 Buckminster Fuller as Captain of Spaceship Earth > 68
6 The Ecological Colonization of Space > 83
7 Taking Ground Control of Spaceship Earth > 96
8 The Closed World of Ecological Architecture > 113
Conclusion: The Unification of Art and Science > 126

CAST OF CHARACTERS > 133

NOTES > 139

INDEX > 177

Illustrations follow p. 82

ACKNOWLEDGMENTS

This book originates from various guest appearances in Hashim Sarkis's Green Modern course at the Graduate School of Design, Harvard University. Students in the Ecology and the Human Condition seminar and in History of Ecology and Environmentalism, which I cotaught with Everett Mendelsohn in the Department of the History of Science at Harvard, were also exposed to the material. I am deeply grateful to Sarkis, Mendelsohn, and all the students who endured the process of turning research into a book. I also extend my gratitude to the generous students of the Science, Culture and Sustainability course at the Center for Development and the Environment, University of Oslo, and to all the graduate students at the Oslo School of Architecture and Design.

Over the years I have had the opportunity to discuss different aspects of this book with Patricia Berman, Robert Brain, Graham Burnett, Jimena Canales, Mark Cioc, Winifred Newman, Antoine Picon, Adam Rome, Tarjei Rønnow, and Lars Svendsen. My wife, Nina Edwards Anker, an architect and the founder of nea studio, has been a key source of inspiration, along with my father, the architect Erik Anker, who has stimulated this project in numerous ways. No words can express my gratitude for their generosity.

My colleagues at the Forum for University History, University of Oslo, offered their time for intellectual discussions. I am particularly grateful to John Peter Collett and Robert Marc Friedman for their patience and support. I am grateful as well to previous colleagues at the

Center for Development and the Environment, University of Oslo. I have also benefited from discussing the entire manuscript with generous students and faculty members in New York University's Environmental Studies Program and the Gallatin School.

At the University of Oslo I had the chance to lure a series of scholars to our Science Studies seminar and other intellectual events. I used the opportunity to engage them in my own interests, which resulted in invaluable discussions with Janet Browne, Peter Galison, Daniel Greenberg, William Clark, Angela Creager, Michael Gordin, Sheila Jasanoff, Matthew Jones, Dan Kevles, Gregg Mitman, Robert Kohler, James Lovelock, Hans-Jörg Rheinberger, Simon Schaffer, James Scott, Steven Shapin, and Sverker Sörlin.

Different, more abbreviated versions of portions of the text have been published as "Buckminster Fuller as Captain of Spaceship Earth," *Minerva* 45 (2007): 417–34; "Graphic Language: Herbert Bayer's Environmental Design," *Environmental History* 12 (2007): 254–79, published by the American Society for Environmental History and the Forest History Society, Durham, NC; "The Closed World of Ecological Architecture," *Journal of Architecture* 10 (2005): 527–52; "The Bauhaus of Nature," *Modernism/Modernity* 12 (2005): 229–51; and "The Ecological Colonization of Space," *Environmental History* 10 (2005): 239–68, published by the American Society for Environmental History and the Forest History Society, Durham, NC. I thank the editors for permission to use the material that appears here.

My thanks also goes to Joseph B. Powell, my excellent editor at Louisiana State University Press, and to two anonymous reviewers of the manuscript. I am also grateful to my first-rate copyeditor, Joanne Allen, and to my assistant, Frøydis Brekken Elvik, who checked all the quotations.

Most of the research for this book was done at Columbia University's outstanding Avery Library. I am indebted to its excellent staff, including Claudia Funke, and to the generosity of the Norwegian Research Council for making the visit possible through grant 148068/V20. My gratitude also goes to the Patrick Geddes Legacy in Oslo and to the Norwegian Non-Fiction Writers and Translators Association for funding.

From Bauhaus to Ecohouse

INTRODUCTION

Global warming has brought ecological design to the forefront of recent architectural journalism and academic debate. Despite claims to novelty, much of this discussion reflects back on earlier ideas. These largely forgotten antecedents deserve notice among practitioners and students of design as sources of inspiration. Some aspects of the history of ecological design, on the other hand, are perhaps not worth emulating. This book describes both. My guiding principle has been to follow the historical relationship of design and the ecological sciences, a relationship that has received little attention until now. How did ecologists and designers engage each other, and what kind of projects did they support?

Starting with attempts to unify design and ecology among modernists in the 1930s, this study clarifies ecological design's mixed history up to the end of the cold war in the late 1980s. It is a history that starts with attempts to relaunch the Bauhaus school among environmentally concerned Britons in London and ends in visions about ecological world management on the part of equally concerned Americans in the United States. The key to this diverse history of both designers and ecologists is their shared attempt to unify art and science in order to find practical solutions to environmental problems.

All historians come to history with their own personal biases and perspectives, and I am no exception. A believer in Freudian theories may find comfort in learning that my father was an eminent architect and that my wife too is a well-known practitioner in the field. Yet for all their influence on me, it was my experience with a particular building that started me wondering about the relationship between ecology and design. Many years ago my parents flew me and my sister from Oslo, where we grew up, to Orlando, Florida, to visit Disney's new Epcot Center. It was very much a family event, and quite exotic for us, coming all the way from Norway. After passing through the gates, we went straight for the line leading into a huge globe named Spaceship Earth, which my father told me had been inspired by the famous architect Richard Buckminster Fuller. Inside we were placed in a visitors' trolley that took us on a ride that featured historic panoramas leading up to futuristic displays of the Experimental Prototype Community of Tomorrow (EPCOT). According to Disney, the future held a world in environmental harmony, with a fully mechanized agriculture and people communicating mostly through machines. To me, Spaceship Earth looked more like the science-fiction spaceships I had seen on film than the Earth I knew. I remember thinking that it surely did not belong to my future. As will be shown in chapters 5 and 6, Disney's visions for the future emerged in lieu of research into ecological engineering and architecture. This book is an attempt to come to terms with this ecological line of reasoning, and it proceeds according to my own trajectory from Europe to the United States.

In 1937 the famous designer László Moholy-Nagy moved from London to Chicago. His work may serve as an introduction to the kind of relationships between ecological architecture and science that I investigate here. Late in life Moholy-Nagy complained that the original meaning of Louis Sullivan's celebrated motto "Form follows function" had been "blurred" to a "cheap commercial slogan," so that its original meaning was lost. According to Moholy-Nagy, the motto should be understood in view of "phenomena occurring in nature," where every form emerges from its proper function.[1] As professor of the metal workshop and responsible for teaching the preliminary design course at the Bauhaus, Weimar, Moholy-Nagy is a key figure in the history of modernist design. In what way was he inspired by science? As will be argued in chapter 1, he was one of many avant-garde modernists who turned their attention to biological sciences to determine functionality and thereby human environmental well-being. Many scientists were enthused by the prospect of improving human evolutionary fitness as well as environmental living conditions through ecological design. One of them was the biologist and ecologist Julian Huxley. His friendship with the founder of the Bauhaus school, Walter Gropius, went back to the early 1930s, a period in which many biologists turned their attention to avant-garde modernist architecture. They saw in this design a way of improving social health and the environment. The designers, on the other hand, took an interest in the biological sciences, as true functionalism to them was a matter of designing according to the laws of nature. This, at least, was on their agenda when they moved to the United States in the late 1930s.

The importance of science to the Bauhaus was noteworthy while the school was active in Germany. Yet the point of departure for this study is events that took place after its faculty moved to London in the mid-1930s and subsequently to the United States. During the London years, biological sciences and ecological concerns came to the attention of the school's former faculty and associated designers, who as a result became deeply concerned about environmental issues. The London ecologists saw in their modernist design possible solutions to human and environmental problems. This environmental thinking and design resurfaced in the United States after most of the school's faculty members moved there in the late 1930s. They influenced the thinking of a new generation of modernist artists and designers, who turned to the biological sciences to find practical solutions to social and environmental ills. That there was a productive unity between architecture and science was a novel and controversial idea among avant-garde designers in the United States, at least if one is to judge from Frank Lloyd Wright's *Testament* (1957). Wright advised his admirers to rely on their own intuition and keep scientists at arm's length.[2]

My focus in this book on European and North American events does not mean that ecological design projects took place only in these regions. Modernist architects also tried to unite design and science in dealing with tropical climate and environments, especially within the British Empire.[3] Also worth mentioning are South American designers such as Roberto Brule Marx, whose "tropical modernism," as Nancy Leys Stephan has shown, represents a Brazilian attempt to fuse design with ecological sciences and sensitivity.[4]

The name Bauhaus needs some explanation, as art and architectural historians tend to associate it strictly with the name of the school that operated in Germany from 1919 to 1933. I use it in a much more inclusive way to also signify the school of thought of its faculty members after the institution was closed by the Nazis. This is not inapt, as it was a name they actively used to describe themselves both in London and in the United States. Though some were inspired by biology while the school was active in Germany, I argue that the fusion of biological reasoning in Bauhaus design took place during the process of trying to reestablish the school in London. This London interlude is largely ignored by art historians,[5] yet the period was important for the school's development in terms of ecological reasoning. Since this fusion of ecological thinking and design is the focus of this study, the life and work of some key Bauhaus faculty members, including Hannes Mayer, Ludwig Mies van der Rohe, and Josef Albers, are not covered. On the other hand, the work of the lesser-known member Herbert Bayer is discussed in some detail. This book is an effort to understand the importance of science to Bauhaus design, a topic that hardly dominates historical studies of the school and its heritage.[6]

The Bauhaus designers were concerned about functionalism in terms of the full human being, including rational, emotional, and environmental needs. By setting the human condition at the center of their projects, they came to guard the humanist tradition. Their adaptation of the ecological sciences illustrates that the humanist legacy, derogatorily labeled "anthropocentrism" by some, has offered more to the history of environmental debate than the chauvinism toward nature that critics point out.[7] This history of ecological design thus illustrates the difficulty of holding up linear histories of "Arcadian" and "managerial" or "industrial" traditions in ecological thinking.[8] Instead of such false dichotomies, the Bauhaus designers illustrate the importance of humanism to environmentalism as it was pursued in their work.

The ways in which Bauhaus modernists came to adapt ecological reasoning and how the scientists responded are the subject of the first four chapters of this study. The remaining chapters focus on how the Bauhaus heritage of trying to unite art and science came to frame the work and thinking of subsequent ecological designers. With the cold war came a radical change in the science of ecology, a shift that dramatically affected the way designers sought to determine functionality. While functionality during the London years was determined by biological investigations into human well-being in the environment, during the postwar years the focus was on human functioning within closed ecosystems. This shift of focus, resulting from new developments in the science of ecological engineering, implied a problematic turn from humanistic toward biocentric designs.

New developments in ecological engineering would frame the work of the generation of architects and designers. Some of them, including the landscape designer Ian McHarg, were trained in the Bauhaus tradition of attempting to unite art and science. That architecture should be based on science came to represent the chief continuation of the Bauhaus legacy, as in the work of Buckminster Fuller. Under headings such as *international style, modernist,* or simply *contemporary design* this new generation of architects continued to determine functionality with the help of science and engineering. While the aesthetic language would change over time, the appeal to science would remain at the core of their work. This was especially the case among designers inspired by the ecological sciences. To understand changes in ecological architecture, one therefore has to investigate key transformations and developments in ecological research. Chapter 6 reviews in some detail the new science of ecological engineering, and chapters 7 and 8 illustrate how this science came to dominate the ways designers sought to handle the environmental crisis.

It may come as a surprise to architects and landscape designers known for trying to design with nature on the ground that the main object of ecological engineering was to discover how to colonize outer space. To construct closed ecological cabins and spaceships for astronauts was the engineers' main task. The significance of this space ecology is at the center of chapters 6–8, which contain a critical discussion of ecological design, in comparison with the first chapters of the book, which are largely supportive. At the heart of this criticism of biocentric designs of the cold war lies a concern for the loss of humanism. The point of departure is the use of colonial terminology in space travel, which was deliberate and in line with the imperial tradition from which ecology as a science emerged.[9] The term *space colony* (instead of, say, *a home in outer space*) was unproblematic, according to Stewart Brand, a leading defender of space colonization, since "no Space natives [were] being colonized."[10] Yet when space colonies became the model for design projects on Spaceship Earth, I argue, all human beings became "Space natives" colonized by ecological reasoning. Human social, political, moral, and historical space was invaded by ecological science aimed at reordering ill-treated human environments according to the managerial ideals of the astronaut's life in the space colony.

The imagined and real environments in space became key sources of inspiration for designers trying to build environmentally friendly landscapes and buildings on Earth. As this study shows, living in harmony with Earth's ecosystem became for the majority of ecological designers a question of adopting space technologies, analytical tools, and ways of living. Their aim was not only to improve life on Earth but also to design an escape from industrial society. As argued in chapter 8, many ecological architects believed that industrial society was doomed and that their task was to design bio-shelters or eco-arks modeled on space cabins in which one could survive if (or rather when) Earth turned into a dead planet like Mars. The social life within these buildings, monitored and managed by the scientifically minded ecologist, was envisioned as resembling the technologically informed lifestyle of astronauts. Life in a future ecologically designed world would focus on biological survival and ecosystem adaptation at the expense of wider cultural, aesthetic, and social values of the humanist legacy. These architectural attempts to mirror closed ecosystems within space cabins resulted in designs construed as enclosed microcosms of the living world.

The colonialist agenda of space research invites the use of postcolonial methodology in the last chapters of this book. Though hardly novel in other areas of historical research, postcolonial analysis has yet to be applied to the history of ecological design.[11] The majority of histories of environmentalism in the 1960s and 1970s have instead been written by a generation of scholars who had the advantage of witnessing events, but not always with the privilege of historical distance. As a result, the prevailing views often entail heroic narratives of environmental champions defending the earth. Ecological design is inspired by a biologically informed vision of humankind embedded in an Arcadian dream of building in harmony with nature, according to its admirers, who do not draw connections to space exploration.[12]

This book is not only for historians of design and architecture but also for historians of science and the environment. Historians of environmentalism have often focused on issues related to the protection of wilderness, an idea that by definition stands in contrast to designed landscapes. As a consequence, the surprisingly rich history of designing objects, buildings, and landscapes in tune with ecological knowledge has often been ignored. The connection between ecological colonization of outer and earthly space has largely been overlooked for the same reason.[13] The few historical analyses of space ecology that do exist have hardly paid attention to its importance to ecologists' understanding of design on Spaceship Earth.[14] Scholars have rightly emphasized the significance of the modeling of closed ecosystems, but they have not placed this methodology in the context of the ecological colonization of space.[15] I argue that the ecological perspective from space had a significant impact on a design on Earth that sought to create what one of its proponents described as a "neo-biological civilization" at the expense of the humanist legacy.[16]

Historians of science have only in the last decade begun to clarify the relationship between architecture and science. One of the first to address the issue was the famous scientist and historian John D. Bernal, who in a 1949 publication discussed the importance of the history of science to architects' changing ideas about functionalism.[17] Bernal was in many ways an exception. Only much later has a growing group of architectural and science historians begun to discuss the relevance of design to science as well as the relevance of science to design. Anthologies such as *The Architecture of Science*, edited by Peter Galison and Emily Thompson (1999), and *Architecture and the Sciences*, edited by Antoine Picon and Alessandra Ponte (2003), have documented a rich field of mutual inspiration between scientists and practicing architects. Perhaps the most compelling arguments have been made in studies by Sylvia Lavin, Christine Macy, Sarah Bonnemaison, and Hadas Steiner. Lavin argues that the work of the architect Richard Neutra is best understood in the context of the science of psychology and the psychoanalytic culture of pleasure in which Neutra and his clients participated. Macy and Bonnemaison document the importance of environmental thinking to architecture. And Steiner focuses on the role of science in reshaping the modernist project.[18]

This book holds that the manner of thinking about the environment can be understood through the chief material manifestation of human agency, design. This is contrary to the view among environmental historians that nature somehow generates social behavior and architecture.[19] The proximity and interrelationship of art, architecture, and the biological sciences are at the core of this book, which holds that views on the household of nature are directly related to buildings.

The history of ecological philosophers may illustrate this connection, as their way of thinking time and again reflects the homes in which they worked. Nature writers were often inspired by the primitive hut, a tradition dating back to the romantics and earlier. Foremost was Jean-Jacques Rousseau, a keen admirer of the primitive hut, as was the Arcadian British nature writer Gilbert White.[20] Similarly, John Muir during his Yosemite years fashioned himself as a bird by living in a "hang-nest" cabin overlooking the environment he admired.[21] The urge for solitary life in a cabin can also be found in the nature writings of Henry David Thoreau, Martin Heidegger, Aldo Leopold, and Arne Næss. Their cabins were located as far as possible from the social realm yet close enough to suggest various schemes for management of the household of nature and society.[22] More recently, the American nature guardian Edward Abbey chose to retreat from the human realm to live parts of his life in a fifty-foot-high fire watchtower at Aztec Peak in Arizona, where he wrote passionately about overlooking nature.[23] The way these philosophers described nature was intrinsically linked to the architecture of their respective cabins. The manner of thinking about the household of nature reflects the house in which the philosopher of nature lived. Given the influence of these thinkers, it is not surprising that life in primitive huts and cabins has inspired environmentalists.[24]

Finally, a history of the ecological sciences and design is an interdisciplinary history. It is therefore worth noting that the word *ecology* here points to both science and architecture. On the one hand, ecology is widely hailed among environmental thinkers as a view of the world in which humans are not at the center; on the other hand, its etymological origin is in the most anthropocentric object on earth, namely, the human house, *oikos*.[25] This raises the question how architectural language traveled into the world of ecological science and back.

1 THE BAUHAUS OF NATURE

In 1937 the biologist and ecologist Julian Huxley hosted a sumptuous farewell dinner party for Walter Gropius upon the occasion of the latter's departure from London to become head of the Harvard School of Design. This major event took place at the fashionable Trocadero, on Oxford Street, with a guestlist that reads like a who's who of modernist design in England.[1] Strangely, among the guests one also finds prominent ecological scientists and environmentalists, which raises the question why *they* were invited to the festivity. Ecologists would seem to be unlikely guests at a party in honor of a Bauhaus architect. Social gatherings are often telling indications of an intellectual climate, however, and Gropius's farewell dinner was no exception. What brought Bauhaus designers and ecologists together, I will argue, was a shared belief that the human household should be modeled on the household of nature.

THE LONDON BAUHAUS

The arrival of former Bauhaus faculty members in London energized the city's designers and intellectuals. After fleeing from Nazi harassment, Walter Gropius (who arrived in 1934 and left in 1937), Marcel Breuer, László Moholy-Nagy (both of whom were there from 1935 to 1937), and Herbert Bayer (who visited briefly in 1937) were able to meet again as a group, something they had been unable to do since Gropius's 1928 resignation from the Dessau school in Germany. They set forth to reestablish the Bauhaus school in London.[2] The guestlist for Gropius's farewell dinner is an indication of who responded favorably to their ambition. Those attending the party may be labeled the "London Bauhaus." What brought the group of designers, town planners, and environmentalists together was a shared belief that Bauhaus design could solve social as well as environmental problems.

The former Bauhaus faculty settled in the Hampstead section of London, which at the time was a community of avant-garde designers, intellectuals, and artists. They moved into a brand-new apartment complex, the Lawn Road Flats—also known as the Isokon Building—the first modernist residence in London. Designed by Wells Coates, the building featured a common room, and there were laundry, cleaning, meal, and garage services. From his window Moholy-Nagy could enjoy a garden of "only trees, which is very peaceful, especially in London."[3]

The list of carefully selected tenants included a host of left-leaning intellectuals and designers, who enjoyed what Gropius described as "an exciting housing laboratory, both socially and technically." Technically, the building was to be a true machine for living, with state-of-the-art furniture and novelties like built-in cooking and washing facilities. Socially, the apartment complex was to promote collective life and liberate the tenants from the burden of personal possessions.[4] Both Moholy-Nagy and Gropius suffered from the language barrier; the latter spoke only "three words of English."[5] Yet they were able to overcome the obstacle thanks to the communal spirit of the Flats. The tenants were encouraged to nurture a fellowship modeled on the Bauhaus workshop, and the school's faculty were recruited as tenants to secure the intellectual climate for what in effect was a socialist architectural experiment. The building was to function like a park, where people could come and go, and it quickly became a hub for the promotion of Bauhaus design.

As the building's architect, Coates was in the midst of gatherings that soon evolved into the Modern Architectural Research Group, or MARS. This group had about sixty members and included notable designers such as Moholy-Nagy, Gropius, Maxwell Fry (who collaborated with Gropius on several projects), Serge Chermayeff, Morton Shand, Godfrey Samuel, John Gloag, and the Russian émigré Berthold Lubetkin (who arrived in London in 1930), as well as the young Danish engineer Ove Arup (who was a frequent visitor at the Lawn Road Flats).[6]

They would all meet in the common room, known from 1937 as the Isobar Club, which was designed by Breuer. During his London years Breuer temporary abandoned his trademark tubular steel furniture and began to experiment with plywood. This was very much in the spirit of the debates at the Lawn Road Flats, which focused on ways in which one could build in harmony with the human body and incorporate organic materials and processes in furniture construction. The result was a series of wood-only tables and chairs known as the Isokon Laminated Furniture series, produced by Jack Pritchard. It includes Breuer's famous masterpiece of modernist furniture, the Isokon Chaise Longue Chair of 1936.

The brothers Aladar and Victor Olgyay were two visitors who enjoyed the Isobar Club with Breuer. They met in London in 1937, the year they completed their Városmajor Utca housing project in Budapest, which immediately caught Breuer's attention. The housing project embraced the garden and rejected the street:

> The most essential characteristic about this house is its preferring not to face the street. Not because this particular street in Buda would be distasteful. On the contrary, it is attractive. No architect would have dared twenty or thirty years ago to build a house in which the inhabitants would have no view of the street. The street is the living artery of the town. Why should we hate it? But the street represents noise, dust and distraction; why—we ask today—should the house be overwhelmed by it, with our leisure disturbed, and even our breathing poisoned by infected air? [Instead they] planned a reverse house backing the street, facing completely the garden.... Every room faces the hill and the big trees of the garden. ... every flat can enjoy the garden air.[7]

The Olgyay brothers would later flee from Budapest to New York, where they would explore, among other things, solar control and shading devices in an attempt to develop bioclimatic design.[8]

Their rejection of the street and embracement of the garden represents the kind of issues debated by MARS members at the Isobar Club in the 1930s in discussions that generally focused on the role of biology in the reshaping of society. The MARS group became advocates of environmental sensitivity: "There must be no antagonism between architecture and its natural setting," they pointed out in their exhibition manifesto of 1938. A drawing of a tree growing through a building was to illustrate that "the architecture of the house embraces the garden. House and garden coalesce, a single unit in the landscape."[9] Or as Gropius pointed out, "The utilization of flat roofs as 'grounds' offers us a means of re-acclimatizing nature amidst the stony deserts of our great towns; for the plots from which she has been evicted to make room for buildings can be given back to her up aloft. Seen from the skies, the leafy house-tops of the cities of the future will look like endless chains of hanging gardens."[10]

This appeal reflected values and ideas promoted by environmentalists such as Clough Williams-Ellis, who thought modernist design could save Britain from ecological destruction. Later hailed by Lewis Mumford as the founder of the ecological movement, Williams-Ellis became known for his *England and the Octopus*, of 1928, in which he rages against the evils of aesthetic and physical pollution of the countryside.[11] Along with Patrick Abercrombie, professor of town planning at London University, Williams-Ellis was on a crusade against unregulated development of the English landscape. Both he and John Summerson, with whom he coauthored *Architecture Here and Now*, had a progressive view of history. For them, modern architecture, with its focus on light and fresh air, represented advancement in public health as well as a remedy that could halt environmental destructions of past developments. As the historian Paul Overy has shown, there was a general effort among modernists to design buildings so that they provided light, air, and openness and thus health to their users.[12] Bauhaus design was particularly promising because it hailed a regeneration of the craftwork Williams-Ellis associated with the traditional English cottage.[13] This environmentalist group championed, as the historian David Matless has argued, a modernist ecological order and buildings, towns, and landscapes that benefited the people who lived in them. What they feared was individualism, laissez-faire development, and rustic nostalgia for the past. They were all welcomed by the MARS group as visitors at Lawn Road Flats, and all of them were on the guestlist for Gropius's farewell party.

Another meeting place for the Bauhaus devotees was the residence of H. G. Wells, who also lived in Hampstead. He was one of the most famous writers of popular science, novels, and science fiction of his time, arguing that the human condition should be understood from an ecological point of view. From the late 1920s he used "human ecology" as his chief methodological tool. His home was known for what the bourgeois would call a salon, though his left-wing friends referred to them as meetings. Wells was a familiar figure in Soviet circles in London, where he engaged with Lubetkin as well as the Russian-born architect Chermayeff. It was through Wells's secretary, Moura Budberg, that Wells also came to know the Hungarian film producer Alexander Korda, who expressed a desire to make a film based on Wells's socialist ideas about architecture.[14] One of the key debating points in their gatherings was the importance of evolutionary biology and ecology in understanding the history and future welfare of the working class and the environment. These ideas became particularly important to Chermayeff, who after moving to the United States became dedicated to environmental design.

The chief source of inspiration for Wells's interest in ecology was Julian Huxley. As secretary of the Zoological Society, Huxley enjoyed a spacious residence at the London Zoo, which he had made into a showroom for modernist design. Here scientists, architects, urban planners, and the environmentalist circle around Williams-Ellis met for discussions. The group, also present at the Gropius farewell party, included Charles Herbert Reilly, William G. Holford, Conrad H. Waddington, Eric L. Bird, Edward Max Nicholson, and James E. R. McDonagh. Their program for saving mankind from environmental, economic, and social destruction through scientific planning found its voice in Political and Economic Planning Organisation, or PEP (see chapter 2).

LEARNING FROM NATURE'S WORKSHOP

The London writings of Moholy-Nagy may provide a preliminary window into the debates about the evolutionary survival of human species that took place at the Lawn Road Flats. There are only sparse discussions of his biological outlook in the literature about the Bauhaus, even though he stressed the importance of biology in his work and his educational program. The architectural historian Reyner Banham, for example, focuses almost exclusively on the importance of mechanization to modernist design in general and to Moholy-Nagy in particular, as does Kristina Passuth, Moholy-Nagy's biographer, who also argues that Moholy-Nagy's "activities became highly disparate, even to the point of fragmentation" during his London years.[15] What might look like fragmentation was actually a relaunching of the Bauhaus as an ecologically inspired program of design.

The Hungarian-born Moholy-Nagy had been a professor in the metal workshop and responsible for teaching the preliminary design course at the Bauhaus in Weimar. He and Gropius had previously compiled a series of books about Bauhaus design that included his own *Von Material zu Arkitektur* (1929) in English under the title *The New Vision* (1930). For many English-speaking designers *The New Vision* provided their first encounter with Bauhaus research methods, and according to the architectural historian John Summerson, it was one of the few books that contained a comprehensive theory of modern architecture.[16]

In *The New Vision* Moholy-Nagy advised his readers to use "nature as a constructional model" and always look for "prototypes in nature" to determine functionality.[17] *Functionalism* is a key word in the book. Moholy-Nagy believed that the future held a new harmony between humans and their earthly environment if forms in design followed biological functions. Nature's evolutionary development was analogous to the development of an individual organism, he believed. This belief was based on the assumption that the evolutionary history of species in nature (the phylogeny) was recapitulated in the development of a human being (the ontogeny). It was consequently important to understand the processes in nature in order to determine the functionality of design for human beings. Functionalist design was a matter of saving society from the degeneration and criminality associated with traditional ornamental arts.[18]

Humans were governed by their biological nature. Consequently artifacts would only be functional if they were construed in relation to human biology, Moholy-Nagy explained in *The New Vision*. Bauhaus design was to "guarantee an organic development" by following the laws of nature; thus it was to allow its users to "follow biological rhythms" so that people's daily "lives would be less hysterical and less empty."[19] That is why Moholy-Nagy stressed the importance of "striving toward those timeless biological fundamentals of expression" that could capture a person's full human potential as an "integrated" being in terms of "biological functions" in "natural balance" with his or her "intellectual and emotional power."[20] The task of design was to create a culture that strengthened people's ability to function biologically, and the way to achieve this end was to make sure that design mirrored the balance of nature. "Technical progress should never be the goal, but instead the means" for a healthy biological life, he argued.[21] The biological welfare of the individual as an actor within the larger matrix of society and the larger environment was equally important. "Art, science, technology, education, [and] politics," Moholy-Nagy argued, were all disciplines that contributed to the "rational safeguarding of organic, biologically conditioned functions" of society and the environment.[22]

The chief source of inspiration for this design program was the Hungarian biologist Raoul H. Francé. Though Francé is largely forgotten today, in the interwar period he was a bestselling author and director of the prestigious Biological Institute of the German Micrological Society in Munich. He was an outspoken defender of psychobiology, the theory that a certain dynamic psyche in living matter is a driving force in evolution. As one of the founders of soil ecology, he argued that the earth had a dynamic power that gave plants a psychic energy whose goal was evolutionary harmony among living organisms. Humans could benefit from the earth's vital powers, Francé argued, if they learned to copy nature's inventions. For example, he designed a salt shaker based on a plant's technique for distributing seeds.

Toward the end of copying nature for the benefit of humans, Francé created *bio-technique,* the science of bionics. The aim of bionics was to study nature's workshop to generate principles, techniques, and processes that could be applied to human technologies so that human society would live in harmony with nature. The structures of plants and of their biotic communities should serve as models for architecture and city planning, respectively. It was possible to imagine a "futuristic utopia," Francé argued, "if the Doctrine of Life would become the load-star of human institutions" and "the optimally functional form" of plants and plant communities was applied to the development of new technology, design, architecture, and urban planning."[23] If designers and architects only understood and learned from nature's own workshop, Francé argued in his futuristic pamphlets, humans would live in health and peace not only among themselves but also with the earth. One merely had to learn from nature's workshop to find out what humans should do, he claimed in *Die Pflanze als Erfinder* (Plants as Inventors), of 1920, a book frequently quoted by Moholy-Nagy.

Moholy-Nagy was not the only designer enthused by Francé's thinking. Both Siegfried Ebeling and Ernst May used bio-technique in their design. May, for example, followed the method when designing housing at Frankfurt-Römerstadt (1926). Bio-technique and the housing project later inspired Lewis Mumford's visions of the future urban order for human culture.[24]

The overall aim of Moholy-Nagy's research was to find a design method that would set human life in harmony with nature's economy as understood by Francé. In his artwork he investigated space relationships and functionality in relation to biological needs. He defined architecture "as an organic component of living" and argued that "architecture will be brought to its fullest realization only when the deepest knowledge of human life in the biological whole is available."[25] While living in London he drew up ambitious plans for a space modulator (not to be confused with his famous light modulator of 1930) in the form of a kaleidoscope that could produce every possible space relationship. The aim was to research "the biological bases of space experience" so that the architect and the designer could proportionally transfer the spatial order of nature into the human realm.[26] The numerous drawings Moholy-Nagy generated from this research should thus be read as attempts to understand how human biology functions in different types of spaces.

Bauhaus design was to reconcile the artificial and the natural in a way that would both enhance human life potentials and create a harmonious environment. Moholy-Nagy was pursuing an indirect argument for ecological protection, namely, that a well-functioning biotic community was a precondition for a well-functioning human society. "The new architecture on its highest plane will be called upon to remove the old conflict between organic and artificial, between open and closed, between country and city," Moholy-Nagy told British architects. He was inspired by Francé's idea that every organic object had a harmonic organization that manifested itself internally as a balanced structure and externally in the shape of ecological communities. Moholy-Nagy developed his program of social responsibility accordingly, by providing communities with structurally sound buildings that gave support to human and nonhuman biological needs. "The thesis on which the Bauhaus was built," he argued, referring to Gropius's first public lecture in London, was "that art and architecture which fail to serve for the betterment of our environment are socially destructive by aggravating instead of healing the ills of an inequitable social system."[27] Bauhaus design would determine "the fate of our generation and the next" if it successfully used the biological forces of life to improve social, economic, technical, and hygienic matters, so that society would live in harmony with nature.[28]

Moholy-Nagy also tried to capture the vital life force of evolution outlined by Francé in his photographic art.[29] He saw art as an expression of the dynamic forces of life in matter: one needed to "replace the *static* principle of *classical art* with the *dynamic* principle of universal life."[30] In his abstract film projects, such as *Light-Play: Black-White-Gray* (1932), Moholy-Nagy tried to capture the dynamic life force on the screen.[31] This was also his agenda in his commissioned photo illustrations for *Eton Portrait* (1937), *An Oxford University Chest* (1938), and the guidebook *The Street Markets of London* (1936). In these three picture books Moholy-Nagy used a snapshot technique that captured Eton and Oxford dons as well as market dealers in action. In capturing the moment with snapshots as opposed to posed photographs, Moholy-Nagy saw himself as a scientist "providing a truthful record of objective determined fact."[32] These facts were also evidence of a class-ridden society, shown in images of places where the biological forces failed to reach their potential. At Oxford, for example, he juxtaposed images of well-to-do students and faculty members with images of the poor as if to remind the reader that the creative energy of academic life all too often was a privilege of the rich.

This attempt to capture nature in action also inspired Herbert Bayer, the former head of the printing and advertising workshop at the Bauhaus in Dessau, who visited London in the spring of 1937 for an exhibition of his paintings at the London Gallery.

In London Moholy-Nagy was involved in several science documentary films aimed at visualizing the productive relationship between science, technology, and design.[33] He also made two documentary films that further illustrate his biologically inspired design program. The first, commissioned by a London documentary film company, was *The Life of the Lobster,* released in 1935. It is a sixteen-minute naturalistic documentation of the growth of the lobster from tiny "crawfish" to old age, as well as of the fisherman's struggle to find the lobsters. What possible interest could an artist like Moholy-Nagy have in the life cycle of the lobster? A hint can be found in a 1937 article by him entitled "The New Bauhaus and Space Relationship." Using the horseshoe crab as his example, he explains that its "prehistoric animal shell is constructed in such a wonderful and economical way that we could immediately adapt it to a fine bakelite or other molded plastic form."[34] The horseshoe crab thus served as an example of the usefulness of natural forms as models for human artifacts. The point of the film was thus to show designers and architects alike that they could learn about form and function by observing the forms and life of animals like lobsters.

The second film Moholy-Nagy made in London was a fifteen-minute silent documentary commissioned by Alfred Barr, the director at the Museum of Modern Art, entitled *The New Architecture of the London Zoo* (1936). It was an attempt to document the space relationships within a building, and it was done in roughly the same fashion as the space-modulator project described above. "I protested against such a naturalistic approach," the architect of the buildings, Berthold Lubetkin, commented, since the film apparently aimed "simply to record, and maintained that the world was full of new shapes, textures and movements."[35] Lubetkin had clearly expected something along the lines of Moholy-Nagy's use of light and shadows in his pre-London films, though the naturalistic language of the film fits well with Moholy-Nagy's project of documenting different biological experiences of space that humans share with animals. Lubetkin's comment also reflected a philosophical difference between the vitalist-informed design Moholy-Nagy advocated and the geometric or mechanistic biology upon which Lubetkin's design was based.

FROM ANIMAL HOUSE TO BAUHAUS

In the summer of 1934 the London Zoo opened with fanfare its new penguin pool, the culmination of a series of new buildings designed according to the Bauhaus style. It was immediately recognized as a masterpiece of the avant-garde, and its famous double-helix ramps have been a pilgrimage site for admirers of modernist architecture ever since.

How did such design relate to ecology, animal welfare, and penguins? The zoo staff used Bauhaus design to promote a socialist-inspired view of the connection between animal and human nature. They saw a close connection between animals and humans and consequently an evolutionary development from the animal house to the Bauhaus, which offered health, welfare, and peaceful relationships between humans and the natural world.

The pool was designed by the Tecton Company, led by Lubetkin, with engineering support from Ove Arup. Lubetkin, who for a while shared lodgings with the Marxist physicist John D. Bernal, was a political revolutionary who jumped at the opportunity to display modern architecture to the masses. Like Le Corbusier, Lubetkin believed that geometric forms were the fundamental building blocks of nature. Though he disagreed with Moholy-Nagy's vitalist biological outlook, he shared his opinion that forms of nature ought to be the models for functionalist design. Moreover, he argued that "there are two possible methods of approach to the problem of zoo design; the first, which may be called the 'naturalistic' method, is typified in the Hamburg and Paris zoos, where an attempt is made, as far as possible, to reproduce the natural habitat of each animal; the second approach, which for want of a better word, we may call the 'geometric,' consists of designing architectural settings for the animals in such a way as to present them dramatically to the public, in an atmosphere comparable to that of a circus."[36] Animals had long played an important part in the Russian circus, so it is not surprising that Lubetkin evoked this idea as a model in defense of his "geometric" approach.

The architectural historians of the London Zoo have also understood its architecture as an example of presenting nature as a circus for human entertainment.[37] Architectural reviews at the time reinforced the image of the pool as a circus for human amusement. The *Architectural Review*, for example, described the "theatrical quality" of the pool as "a suitable setting for any latent publicity talent" among the penguins.[38] Similarly, in *Architect and Building News* a reviewer marveled at how penguins with "a taste for publicity" enjoyed themselves.[39] A reviewer in *Architects' Journal* argued that the pool was a genuine attempt "to preserve the birds from the boredom which generally overtakes all zoo inhabitants," since the birds now had an opportunity to display their social talents on the double-helix ramps.[40] These views were not taken lightly by an American zoo critic who argued that the London Zoo projects had the "flavor of a circus or a country carnival" for "pure pleasure and amusement of their owners, with the tragic result that "the educational or scientific value of an English zoo is nearly zero."[41]

The problem with the interpretation of the zoo as a circus by historians today as well as reviewers of the 1930s is that it does not explain why the keepers of the zoo chose to house their animals in such an environment. Having been trained as biologists, they fashioned themselves as serious scientists and not as managers of a zoological circus. If one looks at the Tecton buildings from these scientists' perspective, a quite different view of the roles of zoo architecture and animals emerges. They were the ones who paid the construction bill, and as clients of Tecton they had a significant say in the design of new zoo buildings. Though they welcomed entertainment that could generate general interest in biology (and money from entrance fees), they were not willing to pursue amusement at the expense of their scientific integrity or the welfare of the animals.

The zoo managers believed in a difference of degree and not of kind between animals and humans. The life and welfare of animals within modernist architecture was not a subject for human amusement but served instead as a model promoting public health. It is telling that when the research anatomist at the Zoological Society, Solly Zuckerman, recalls the Tecton architecture firm, his first memories are of yogurt.[42] The seven young Tecton architects, in their mid-twenties, were struggling to find work in a sluggish economy. One of Lubetkin's colleagues tried to support himself by selling his eastern European family's dairy secret, which, incidentally, introduced yogurt to the English market. It was exactly such new, white, clean, and healthful dairy products that intrigued Zuckerman, who was deeply concerned about public health. Politically he was decidedly a leftist and consequently sympathetic to new ideas from eastern Europe, such as healthful, inexpensive yogurt for the masses. Zuckerman first heard about the yogurt from Godfrey Samuel (a MARS member and son of the Liberal politician Herbert Samuel), who had recently graduated from the Architectural Association and now was a member of the Tecton team. Zuckerman had been introduced to Samuel by his friend Philip D'Arcy Hurt, who at the time was researching the alarming spread of tuberculosis in England. He believed that modernist architecture could provide more healthful homes, with better air and more light, for the English poor.[43] Zuckerman might also have learned about Ove Arup through the latter's late father, the veterinarian Johannes Arup, who was well known among British zoologists.[44] The story of how Zuckerman met and eventually came to promote Tecton among the zoologists thus points toward issues of public health rather than elaborate design and circuses.

There was more to the new architecture than public health and yogurt. When the Zoological Society purchased two young gorillas for its zoo in 1932, Zuckerman became the chief authority on how to house them. It was with respect to this problem that he started to promote Tecton. He had just published a major article entitled "The Menstrual Cycle of Primates," in which he argued that the cycle was the key to understanding the sexual and thus the social life of primates.[45] The aim of the design for the gorilla house was not only the sexual health of gorillas; the Zoological Society had both a scientific and an economic interest in seeing a baby gorilla born in Regent's Park. Fresh air and stable temperatures in a gorilla house were apparently vital for regular menstruation, a fact that would determine Tecton's design. According to Zuckerman, the difference between human and animal behavior was "almost certainly one of degree only"; thus the life of primates might be seen as "a crude picture of a social level from which emerged our earliest human ancestors."[46] He argued this in *The Social Life of Monkeys* (1932), in which the life of primates serves as a model for explaining deeper sexual and social instincts in humans. Zoo visitors who observed the gorillas might thus also observe and reflect upon their own primitive desires. It was consequently of moral importance to place the gorillas in a home that was a model for healthful living.

Zuckerman was the first of several scholars at the zoo who were interested in modernist architecture. He introduced Lubetkin to the superintendent, Geoffrey Vevers, and to the secretary of the Zoological Society, Peter Chalmers Mitchell, who also came to promote Bauhaus design. Chalmers Mitchell believed that evolutionary biology gave support to a cooperative model of social behavior and that peaceful coexistence was the best strategy for evolutionary survival among animals as well as humans.[47] All species could prosper if they were given the opportunity to live in a healthy and peaceful environment. Using penguins as an example, he argued that "the most unlikely animals seem to thrive under what would seem the most unnatural conditions," provided that they had "freedom from enemies, regular food and general hygiene."[48] The same would hold for workers and the poor, who were in desperate need of being liberated from their "natural" condition of criminal and filthy slums. It was thus politically important to display thriving animals in an unnatural setting, suggesting that humans too could prosper if given the opportunity to live in a new and peaceful environment. Chalmers Mitchell arranged for unemployed miners to perform a large part of the heavy work in constructing the zoo buildings, since this allowed them to be in "good open-air conditions until they had recovered sufficiently mentally and physically to pass out into the ordinary labour world."[49] Taking both penguins and miners out of their usual environments and placing them in a fresh and peaceful setting proved to him that humans and animals alike could prosper if they were given a second chance. People's and animals' health and welfare were of equal concern, and the new architecture was to promote them. The zoo, he believed, should be a place where humans could have a "close up view" and reflect on the instincts they shared with other animals.[50]

The newspapers and the popular press described the new buildings as healthful environments for animals and humans alike. The *Times*, for example, noted that the penguin pool was to have "architectural unity and pleasing effect, and at the same time be thoroughly hygienic, give the birds what they require, and afford ample space for visitors."[51] Likewise, the gorilla house was praised for being a healthful and hygienic home rather than for its inventive design.[52] A 1936 popular history of the London Zoo described it as a "very good indicator as regards the nation's material, and indeed spiritual health" and said that its new architecture should serve as a model for the "ultra modern human dwelling-house," complete with "access to the outer air" and opportunities for a "sun-bath."[53] A healthful environment at the zoo was to be a model for healthful human living in the future. Indeed, as the historian of architecture Hadas Steiner has shown, the infrastructure and buildings of the London Zoo were imbedded with modernist visions for urban design and planning.[54]

The cooperative view of society and evolution was based on a mechanist footing, Chalmers Mitchell argued in a 1932 lecture entitled "Logic and Law in Biology."[55] Inspired by Marxist materialism, he thought sound social policies should be founded on logic, physics, biology, and social psychology. The aesthetic language of geometric logic in Tecton's penguin pool was to Chalmers Mitchell a visual representation of the foundation of modern biology. These ideas were inspired by new ways of understanding evolutionary biology among left-leaning London scientists. Since the rediscovery of Gregor Mendel's theories in 1900, biologists had fought over whether their research should have an evolutionary or a genetic footing, and a mathematical turn in biological methodology was seen as a revolution in the field.[56] The publication of Ronald A. Fisher's *Genetical Theory of Natural Selection* in 1930 marked an important moment in the modernist turn in biology. It was possible, according to Fisher, to make a synthesis of evolutionary theory and genetics if biology were based on a mathematical foundation. This evolutionary synthesis had its precursor in mechanistic views of nature promoted by John B. S. Haldane and Julian Huxley in their popular book *Animal Biology*, of 1927.[57] Huxley saw the new mechanistic and mathematical approach in biology as a key material basis for a successful, orderly planning of human society with a new urban matrix of mathematically inspired Bauhaus architecture.[58] The geometric order of the zoo buildings became a visual representation of the promising mathematical turn in biological research models. The new Bauhaus households thus mirrored the new order of the household of nature.

In April 1934 the Zoological Society hired Huxley to replace Chalmers Mitchell. While Chalmers Mitchell retired (and went to fight on the socialist side of the Spanish Civil War), Huxley continued the program of avant-garde design at the zoo, which eventually put him in the position of hosting the farewell dinner party for Walter Gropius in 1937.

2 PLANNING THE ECONOMY OF NATURE

Biologists and designers alike thought their disciplines had a prominent role to play in staking out an alternative future to destructive capitalism. In this chapter I discuss some social and environmental concerns shared by scholar and architect friends of Julian Huxley. They were worried about the well-being of humans living in the dark homes and polluted environments of London. Many biologists viewed such living conditions as a matter of evolutionary survival of the human species. They saw the solution to the problems in urban planning, landscape design, and new architecture providing fresh air and sunlight. Their call for new design projects was based on scientific plans to bring a new environmental order to society and nature. They envisioned a future ideal society in harmony with the household of nature.

A NATIONAL PLAN FOR THE ECONOMY OF NATURE

London in the 1930s was the scene of a broad academic movement for the economic and social betterment of Britain and the world. Within this context, ecological reasoning became one venue for social criticism, environmentalism, and hopes of a more eco-friendly society for the future. The economic depression of the early 1930s generated skepticism toward laissez-faire capitalism. The economic downturn and crash in the stock market was often blamed on a social system based on greedy individualism. The result, the critics claimed, was unhygienic slums and poverty among those who for social or economic reasons had been unable to benefit from the economic boom. For many the new Soviet system appeared to be a viable alternative, and social and economic planning of a more feasible political system was on the agenda of progressive and socially concerned intellectuals.

One key advocate of such planning was the Oxford-trained ecologist Max Nicholson, who in 1931 published "A National Plan for Great Britain," aimed at saving the nation from economic depression.[1] The plan became the vehicle for much debate in the *Week-End Review*, edited by Gerald Barry. The Political and Economic Planning Organisation (PEP) was instituted as a result of the debate, and the journal soon became its semi-official publication. Nicholson served as the assistant editor, with responsibility for covering environmental and scientific issues, town and city planning, and architectural debates. A chief focus of this journal was the slums of London. There was an urgent need for more scientific knowledge about the poor and practical research on what to do about them.[2] One path that he found a particularly promising direction was scientific town and environmental planning.[3] Another equally promising path was to research new construction materials and architectural designs in order establish new building techniques and a minimum standard for housing. When the new Building Centre opened in London in 1932, for example, Nicholson marveled at "the breathless development of architecture, of engineering, of new materials and inventions" on display.[4]

Progressive architectural design was one of the National Plan's core aspects, and many architects, especially Walter Gropius, regarded PEP with hope and admiration. As an outspoken socialist and ecologist, Nicholson argued that Londoners faced a serious population and slum crisis and that scientists, architects, and planners offered new housing and urban designs to deal with the problem.[5] For years he had argued that economic planning was the solution to unemployment and a way to control the "large and complex machine" of industrial civilization.[6] Regional planning was the key "for society to engineer its own development as it engineers the services on which it depends."[7] He economically supported a village college in Histon designed by Gropius and Maxwell Fry that aimed at moving people out of cities and into small, presumably healthful country towns. This project developed in response to a manifesto on minimum housing standards signed by a series of modernist architects and published in the *Week-End Review*. It promoted new standards for room size, light, ventilation, sanitation, heating, and access to water, as well as proximity to the workplace, grocery stores, and childcare.[8] The basic idea was that old-fashioned housing design reinforced an unfortunate dualism between humans and nature, while the new Bauhaus architecture promised to reunite humans with nature through healthful living. The goal was to plan cities that could secure the evolutionary survival of the human species.

Two leading members of PEP were Julian Huxley and H. G. Wells, both of whom had been greatly inspired by Nicholson's plan for Great Britain. Ecological sciences and Bauhaus design would merge in their visionary ideas for a planned society in harmony with nature's economy.

ARCHITECTURAL PLANNING AS *IF I WERE DICTATOR*

In his new job, Julian Huxley would continue the program of Bauhaus architecture at the London Zoo with a Studio of Animal Art, designed by Tecton and completed in 1937. It was used daily by about two hundred students who came to draw live animals. Despite the success of the Studio of Animal Art, Huxley tried in vain to persuade the zoo's council to extend the program of modernist design to a new elephant house.

The council rejected proposals by both Fry and Lubetkin for both financial and political reasons. Huxley was hired by the council based chiefly on the public success of his 1934 nature film *The Private Life of the Gannets*, directed by Alexander Korda.[9] As soon as he got the job, he published one of his most high profile socialist books, tellingly entitled *If I Were Dictator* (1934), which did not go over well with conservative members of the council. In effect, it brought the modernist building program at the zoo to a halt.

The book derived from the debates in the *Week-End Review* about the need for a national plan. Huxley would use much ink on the importance of the new architecture to the scientific planning of human biological welfare. He was convinced of the need for a planned economy that could generate massive housing projects, though he was not intellectually satisfied by its methodological foundation and scientific foothold. He believed that such planning required a synthesizing method linking the natural realm, the human mind, and social politics. Huxley set forth to secure this linkage by reducing biology, psychology, and sociology to mechanism and mathematics. Huxley laid out this program in a 1933 cover article entitled "The Biology of Human Nature." The article was introduced to the readers by Nicholson, who explained that "any solution of the world crisis depends on fundamental changes in human outlook" and that Huxley's mechanistic humanism was an attempt to "bridge the gulf between biology, psychology and politics."[10] In the article, Huxley reviews the pathbreaking research by his colleague Ronald A. Fisher on how to synthesize genetics and evolutionary biology through mathematics and on the importance of this evolutionary synthesis for national economic and political planning.

In *If I Were Dictator* Huxley summarized his reasoning for supporting the Political and Economic Planning Organization. The book was scrutinized in its entirety by Nicholson before its publication and became an important summary of popular views within PEP.[11] Huxley's brother Aldous had for a short period joined PEP but soon resigned to write his novel *Brave New World*, which masterfully ridicules the organization as well as Julian Huxley's, Nicholson's, and Wells's various schemes for an ecology-based society.[12] Julian Huxley would not let this damaging critique go unchallenged, and he set out to point out the positive sides of scientific planning. (The title of his book plays on what he thought would happen if he were the scientific dictator imagined in the *Brave New World*.) To get out of economic depression, he argued, it was urgent to halt "*laisser-faire* individualism, for that is not organic," and leave social steering to a board of directors of the economy of nature with a scientist (preferably himself) at the helm.[13]

"A good Dictator" assisted by the intellectual "*élite* of the country," he argued, could pursue the right vision for helping humans achieve a higher level of living and health and thus secure the survival of humankind.[14] Science was to serve social ends, and one key research topic, Huxley argued, was how to generate better and more healthful buildings. Such progressive buildings were to be found in modern design, and not in design of the past. In a letter to the editor of the *Times*, for example, he ridiculed the plans for the New Bodleian Library at Oxford, which were inspired by architecture from the past, so that visitors would "have to look in their guide-books to find out that it was built in the twentieth and not in the sixteenth century."[15] The scandal, as Huxley saw it, was that a research university had not engaged in architectural research when raising a new building.

The kind of research that architects and engineers conducted in designing the Building Research Station at Watford, "applying the methods of mass-production and modern technology to housing" to make possible cheap prefabrication of homes for the poor, was to Huxley exemplary.[16] At the Station scientists investigated the standardization of materials, so that architects and builders could construct new designs based on a reinvented tradition of craftsmanship. They would investigate heating, humidity, noise (such as echoes), and, most important, light, or the art of "sun-planning—siting and designing a house so as to get the maximum amount of sunshine," so that the house of the future would be as hygienic as possible.[17] Such architecture could help fight tuberculosis and other illnesses caused by unhygienic environments. Preventable causes, such as overcrowded or unsanitary buildings and lack of facilities for outdoor recreation, accounted for a great deal of tuberculosis, Huxley believed, and research on new design and architecture could do something about it. The underlying issue was the evolution of culture, which to Huxley only evolved in productive collaboration between imagination and science.

In Huxley's plan, an elite group of scientists elected on the basis of scientific merit would take charge in "the remodelling of the life of a nation."[18] His chief sources of inspiration was the Tennessee Valley Authority, in the United States, and grandiose socialist ideas. Huxley visited Tennessee in 1932 and returned to London deeply impressed by the project's large-scale architectural reorganization of nature and society. It served him as a model for how to proceed in the British Isles. Upon Gropius's departure for the United States, Huxley urged him to visit the region. The architecture of the TVA has been subject to historical analysis, thanks to a first-rate study by Christine Macy and Sarah Bonnemaison.[19] They argue that the design project aimed at establishing a communitarian utopia in which expert planning and cooperation replaced economic competition and conflict. The project, which was part of the New Deal, aimed at restoring a landscape suffering from environmental degradation, as well as helping rural farmers adapt to a modern lifestyle. The chief architect of the project, the Hungarian émigré Roland Wank, evoked the bio-technique methodology of his compatriot Mohol-Nagy by using nature's energy as a model when designing buildings, dams, highways, landscapes, and recreational parks.[20]

In his many public appearances Huxley fashioned himself in the image of the TVA director and architect, conducting "major experiments in social planning" on the British Isles.[21] As master builder of the nation he would offer a New Deal for Britain that abandoned private ownership, cleared slums, and built new housing communities, all in the interest of the needy. In reality, though, Huxley was not a director of his country but only secretary of the London Zoological Society. In this job he was free to project his grand visions into the animal kingdom by making sure that the animals' housing and living conditions served as a model for organizing humans' future world.

Yet Huxley met resistance from fellow radicals who thought the zoo had a petit-bourgeois flavor. It was not clear to all why gorillas and penguins should be better housed than humans living in the London slums. Huxley was of another opinion. The biological difference between humans and animals was one of degree and not of kind; thus animal homes could serve as examples for future human housing. "In housing their animals," Huxley explained, the zoo also had the advantage of being able to "experiment with new and striking architectural design, too advanced for general acceptance outside."[22] Thanks to research in experimental architecture at the zoo, he argued, in the future there would "no longer be the lamentable contrast between the accommodation provided for the gorillas at the London Zoo and the human population of our towns."[23]

THINGS TO COME IN THE NEW BAUHAUS OF NATURE

H. G. Wells was a frequent visitor at the London Zoo. A collaborator with Huxley, he was in the midst of the debates among urban planners, Bauhaus designers, and ecologists about the future welfare of the human and the natural world. A TVA-inspired vision for remodeling the British Isles appears at the end of his science-fiction film *Things to Come* (1936). The architecture imagined in the film was based on an ecological vision of the future that originated in scientific debates in London. The film promoted a new, environmentally responsible society informed by the emergent ecological science of the energy and fluidity of the household of nature. Film historians have described in detail the socialist agenda, the creative use of new design and filming techniques, and the popularity of the film.[24] Yet, the same historians have failed to discuss how the science of ecological engineering informed the film's layout and set design. The final scenes of the film summarize what the London Bauhaus was all about.

Wells's collaboration with Huxley on *The Science of Life* (1930) triggered his interest in ecology. It became one of the most widely read books about the life sciences in the twentieth century. In it they developed a new "Ecological Outlook" on the relationship between the economy of nature and society.[25] This relationship was in serious trouble, they argued, because of the inefficient and extensive use of nature's matter and energy at the expense of future generations. One had to move from free-market capitalism to a planned economy in order to control the "breeding storm" of the human population and the impending environmental havoc.[26] They suggested taking action against smoke pollution, cleaning up the atmosphere by limiting the use of coal and oil, and ecologically sound ways of using fertilizers. Only the progressive evolution of machinery, architecture, and scientific discoveries would solve the ecological crisis. They argued, for example, that "from the point of view of Mr. Everyman's skin," architects should make sure "to keep the air in a room fresh and stimulating" by means of safe ventilation and climate control.[27]

The basic principle behind their way of ordering society was that nature's energy should be channeled into more efficient and thus better human use. Wells thought of an efficient economy as "a branch of ecology; it is the ecology of human species."[28] In his nine-hundred-page treatise about a scientifically planned future, *The Work, Wealth and Happiness of Mankind* (1931), he explains not only architecture but all economic activity in terms of how human animals learn to conquer and control the economy of nature for the benefit of a progressing civilization. Current advances in architecture were signs of evolutionary progress in Wells's ecologically informed outline of human history.[29] For Wells, human ecological adaptability became a guiding tool in judging the success of various human activities and policies.[30] The latest developments in modern architecture were particularly instructive as a vehicle for rebuilding a new world on the ruins of the past. Le Corbusier's *Urbanism,* which Wells read in its English translation as *The City of Tomorrow* (1929), was one of his sources of inspiration.[31] Le Corbusier's focus on the importance of air, sunlight, geometry, and orderly planning was to Wells an example of the importance of architecture in laying out a new ecological world order that would secure the evolutionary survival of humankind. The British architects appreciated the support of modern design from an influential writer like H. G. Wells. He was invited to lecture at the Royal Institute of Architecture, where he argued that "light was becoming a material of architecture."[32]

Architecture was at the heart of Wells's visions of how people of the future would live collectively to effectively control the air and light in buildings. Town and city planning based on ecological zoning of the environment was urgently needed to limit the number of people living in one place and thus bring the population problem under control.[33] A "possible rebuilding of the world," Wells argued, should be the chief task of modern architecture, and he pointed to the work of Ronald Aver Duncan as a promising example.[34] Duncan, a lecturer in the history of architecture at the Architectural Association in London, argued that a historical shift was taking place in construction techniques from wet materials like mud, cement, mortar, and brick to dry materials like steel, plastic and glass as a result of a shift in manufacture from skilled handicraft to mass production. Wells, who had grown up in a poor, wet brick basement, found Duncan's argument about the coming of an age of dry materials compelling because it meant more fresh air and sunlight in buildings.[35] He saw the movement from wet to dry and from handicraft to mass production in architecture as a leap forward in the biological evolution of human development.

In *The Shape of Things to Come* (1933) Wells predicted the rebuilding of society with the help of designers' use of bio-technique and scientific management of the economy of nature.[36] The book received a lot of attention, and a theatrical adaptation of the story subsequently played in London.[37] Wells agreed to let the Hungarian-born film director Alexander Korda direct a film version of the book under the unusual condition that Wells would have control over every aspect of the film.[38] It was the most costly movie of the time because of the large and complex set designs that Wells ordered to illustrate the shape of the future. The result was a saga about the downfall of industrial society through endless wars and environmental problems. At the end of the movie, humanity is saved by a group of engineers, planners, architects, and ecologists who arrive mysteriously in one of Norman Bell Geddes' futuristic planes of 1932.[39] (This sequence was also inspired by Le Corbusier's 1935 book *Aircraft*, which focuses on the importance of aerodynamic design for future societies.)[40] They land and in military formation march out to save Everytown, Wells's image of anyplace on earth. The hero, John Cabal, forms a governing board to steer the world's natural and social economy through enlightened ecological dictatorship.[41] He tells this exclusive group: "We have ideas in common; the freemasonry of efficiency—the brotherhood of science. We are the natural trustees of civilization when everything else has failed."[42] This brotherhood establishes a regime of authoritarian enlightenment founded on principles of ecological reasoning, planning, and architecture.

The story then fast-forwards from the 1930s to 2054. The population problem has solved itself quietly and mysteriously, and a new, fantastic society has arisen. This sequence, which lasts for about ninety seconds, was designed by Moholy-Nagy. It captures the vitalist evolution of humankind from the world of the warlords to the glorious rule of scientific design and management. Using bio-technique, the architects of the future are able to create a better human world by capturing the vital energies of the earth. Engineers build a huge mining machine that blasts its way into the earth and then build a new environment inside an enormous cave. The process of building the cave recapitulates the development of construction techniques from wet to dry materials (following Duncan's theory). A huge electric power plant provides power to a steelworks that transform wet rocks into shining steel plates, from which the new city is built. Indeed, Moholy-Nagy's wife, Sibyl, recalls that her husband tried with "the fantastic technology of the Utopian city of the future . . . [to] eliminate solid form. Houses were no longer obstacles to, but receptacles of, man's natural life force, light."[43] The sequence—accompanied by modern music by Arthur Bliss—is made up of a series of experimental images that capture the evolution of human ecology in action.[44]

The design highlight of the film is an enormous Bauhaus-style cave city below the surface of the old, doomed society and natural world.

In this second world people work in perfect synchronization with machines. Indeed, they are machines. The outlook is ecological: zoologists do not give pet names to the animals they research, because a name entails individualism. Animals are best described in the science of ecology by numbers in the energy flows of an ecosystem. Consequently, the workers in the movie are not given names, only numbers on their chests. Their sole task is housekeeping in the household of nature.

Unfortunately for Moholy-Nagy, Alexander Korda's younger brother Vincent got all the credit for the design in *Things to Come*, since he was listed as "Setting Designer" of the film. It was not until 1938, when Moholy-Nagy published his manifesto article "Why Bauhaus Education?" in the journal *Shelter*, one of the best venues for International Style architecture, that it became clear whose labor was behind Vincent Korda's achievements.[45]

Originally Wells and Alexander Korda had asked Le Corbusier and his colleague Fernand Léger to design the cave city, but they had declined,[46] perhaps because Wells and Le Corbusier did not see eye to eye politically, as Le Corbusier was not a socialist. Alexander Korda had then asked his brother, Vincent, to do the job, along with Wells's son Frank and the special effects director, Ned Mann. Vincent Korda believed in a division of labor between the set designer and the artist,[47] and he therefore delegated much of the architectural design of the cave city to Moholy-Nagy, who for his part had argued for artistic independence.[48] At one point Gropius served as a consultant to Vincent Korda, making sure that the cave would reflect Bauhaus design principles. All of them might also have been inspired by Marcel Breuer, who at the time was busy drawing up his own plan for a garden city that had some of the same features as the cave city in *Things to Come*.[49] The basic plan of the cave, however, came from Ebenezer Howard's garden city, which Wells admired as the city plan of the future.[50]

Wells envisioned a perfectly geometrically ordered ideal ecological society as a sort of island Eden dominated by technology that would channel nature's energy into maximum food production without causing pollution or other environmental problems. In such an Eden people would move on flying walkways, sit on curved balconies, and be transported through transparent ducts, all emphasizing the movement of people and energy in a streamlined ecological city system. Living in a cave meant having full control over all life forces, including light and air. Everytown was built as a closed environmental ecosystem with a perfect balance of air and light securing the best possible living conditions for everyone and thus evolutionary survival for the species. As a pleased ecologist in the film, Passworthy explains to Cabal: "Our light is brighter than the sunshine outside and never before has mankind breathed so sweet an air. We have got the better of nature."[51] A view from a terrace within the cave where the ecological architect, engineer, and planners develop their ideas suggests that the cave also was inspired by Le Corbusier's *Towards a New Architecture,* which had appeared in English in 1927.[52] A series of oxygen machines (trees) on the eleventh floor and the food machines (bushes) on every floor inside the cave secured daily needs. These forests were stacked one on top of the other within the cave to save space (thus anticipating the design of Winy Maas's *Datacity-Metatown* project, of 1999).[53]

This all looks quite fantastic, but to the inhabitants of Everytown living in a cave is perfectly normal. A short conversation between an old man and a little girl looking at some virtual images of the old New York skyline highlights the amusement of youngsters looking back on the days when people lived aboveground:

> "What a funny place New York was—all sticking up and full of windows."
>
> "They built houses like that in the old days."
>
> "Why?"
>
> "They had no light inside their cities as we have. So they had to stick the houses up into the daylight—what there was of it. They had not properly mixed and conditioned air. Everybody lived half out of doors. And windows of soft brittle glass everywhere. The Age of Windows lasted four centuries. They never seemed to realise that we could light the interiors of our houses with sunshine of our own, so that there would be no need to poke our houses up ever so high into the air."
>
> "Weren't the people *tired* going up and down those stairs?"
>
> "They were *all* tired."[54]

Yet some of the citizens of this apparent ecotopia show signs of discontent. A philosopher called Theotocopulos leads a literal underground resistance in questioning scientific progress, modeled after Aldous

Huxley and the questioning of science in *Brave New World* (1932). "What is the good of this Progress?" the philosopher asks in a dramatic speech. "Give the earth peace and leave our human lives alone."[55] In the next scenes there is dramatic tumult, with earthpeace activists trying to stop the launching of the ultimate scientific project of progress: the launching of a "Space Gun" with a bullet-shaped capsule manned by scientists into space. In the final scene the capsule is fired (liquid-fuel rockets were still to come in 1936), leaving a crowd of angry earthpeace romantics on the ground.

What would it be like to live in this ideal ecotopian cave? It is, of course, impossible to tell what life is like in a fictional place and society, yet the recollections of the actors who spent most of the twelve months of shooting in the cave may provide a hint. Cedric Hardwicke, who played Theotocopulos, came to agree with the philosopher's critique of all the progress.[56] The cave was not an ideal world to him, nor to Michael, Korda's nephew, who thought the cave "puritanically neat, clean and dull."[57] Raymond Massey, who played John Cabal, was rather unhappy with a character who he felt spoke in too "large gobs of socialist theory." As he explained, "The picture was fantastically difficult to act. Wells had deliberately formalized the dialogue, . . . we delivered heavy-handed speeches instead of carrying on conversation. Emotion had no place in Wells' new world." Massey, who played the scientific leader of the cave, had no authority on the set, where Wells ruled as a dictator in charge: "We were always the puppets of Wells, completely under his control. Like all socialists, in his forecast of man's future Wells saw nothing but authoritarianism." What was worse, the futuristic costumes he had to wear were extremely uncomfortable, intolerably hot in the summer and freezing cold in the winter: "The costumes gave the proof that the spirit of Marquis de Sade was alive and well and working as a designer for Alex Korda or, rather, for H. G. Wells."[58]

Things to Come received rave reviews in the press. Although it became a box-office hit, it was not a financial success, because of the costly sets. Though some were skeptical about the "mystic communion with machinery" in the society of the future, the Bauhaus design was regarded as "the real triumph" of the film.[59] Architects were also excited about the film and included the design of the cave city in the ongoing debate about urban planning. One architect reviewer suggested that the film pointed to a new configuration of the landscape in which town life would take place in vertical towers and shafts, while horizontal land would be reserved for country life and natural habitats.[60] Another writer in the *Architectural Review* proposed that the cave design was "easily the best work" of functionalism "yet done in England."[61] *Architectural Forum* published a note arguing that the underground city represented "man's final break with his natural environment," a remark that makes sense if one recalls that the cave was to be a second nature based on a new ecological order.[62]

The film was by no means the culmination of Wells's work on ecology. For the rest of his life Wells promoted the urgent need for humans to adapt by means of technology to the ecological reality of life.[63] In 1941, as Britain plunged into war, he published a *Guide to the New World* to cheer up the nation. This was basically the old (2054) vision in a modified (1951) version. Reborn Britain, as he saw it, would be "green as ever," freed from capitalism, with plenty of nature reserves and small high-tech, neatly designed villages in which people lived in harmony with the ecosystem. The housing would be "more of an open air camp," thanks to scientific advances in air-conditioning systems. These accomplishments "will make architecture the most enviable of professions in the world ahead." "Architecture will [in 1951] become the master art, as it was in the days of Pericles," he argued.[64] There would be no slums, and people would live as nomads, moving about with the energy flow of nature.

The *Guide to the New World* was written in the Allies' darkest hour, and Wells thought of his book as providing his fellow countrymen with some hope in a time of despair. Yet disillusion caught up with his utopian visions, as indicated in the title of one of his last novels, *All Aboard for Ararat* (1941).[65] With Europe falling into chaos, he now saw himself as Noah carrying species and scientists in his ark through rough weather to the Promised Land of Ararat, where they would lay the foundation for the new ecological world state.

In reality Gropius and Moholy-Nagy became the Noah figures for the London Bauhaus, taking the remnants of their school to the United States.

3

THE NEW AMERICAN BAUHAUS OF NATURE

In 1938 the Museum of Modern Art in New York introduced the Bauhaus school of thought to the United States.[1] The show was produced by the former Bauhaus faculty members Walter Gropius, László Moholy-Nagy, and Herbert Bayer, who had arrived in the country about a year earlier. In it they presented a vision of a rebuilding of America that was basically a blueprint of ecological design arguments developed during the London years.

WALTER GROPIUS: AN ENVIRONMENTALIST AT HARVARD

When Gropius came to Harvard, he built his new home next to Walden Pond, where Henry Thoreau had once lived. As the American environmental philosopher had done, Gropius used his professorship to warn against industrialism and capitalism, which might dominate human life unless architects approached design and the environment in a responsible way.

This environmental agenda was a dominant theme in Gropius's teachings at Harvard. In his inaugural lecture he emphasized that architectural students needed thorough training in the biological and social sciences in addition to engineering and architectural skills. "Good architecture should be a projection of life itself and that implies an intimate knowledge of biological, social, technical and artistic problems."[2] Gropius would express this ethic in his plea for a design that included all aspects of human life. Design should be capable of "evolving the 'complete being' . . . from his biological centre" so that one could avoid "the rush and convulsion of our mechanical age," he argued.[3] A knowledge of biology was particularly important for an understanding of human nature, the environment, and the most responsible way to build in the landscape. For Gropius, responsibility was a question of nurturing an "organic social structure" by designing with nature and not exploiting it. "Overwhelmed by the miraculous potentialities of the machine," he wrote, "our human greed has interfered with the biological cycle of human companionship which keeps the life of a community healthy."[4] The task of the architect and the urban designer was to save human dwellings and urban cities from the pitfalls of capitalism. "We have to build up carefully an organic-evolutionary procedure for their rehabilitation," he argued.[5] It was important to build parks and plant trees to achieve this goal: "Under trees the urban dweller might restore his troubled soul and find the blessing of a creative pause."[6]

Urban environmental renewal was valuable in itself, according to Gropius, but it should also be seen in terms of protecting the non-urban environment from suburban sprawl. Making cities livable could save the surrounding environment and the larger non-urban habitat from further development. This, at least, was what Gropius told his students at Harvard in the early 1950s: "The greatest responsibility of the planner and architect, I believe, is the protection and development of our habitat. Man has evolved a mutual relationship with nature on earth, but his power to change its surface has grown so tremendously that this may become a curse instead of a blessing. How can we afford to have one beautiful tract of open country after the other bulldozed out of existence, flattened and emptied for the sake of smooth building operations and then filled up by a developer with hundreds of insipid little house units, that will never grow into a community. . . . *Until we love and respect the land almost religiously, its fatal deterioration will go on.*"[7]

This sentiment was widespread not only among members of Gropius's circle. His former colleague Maxwell Fry, for example, would publish environmental architectural and urban designs as late as 1976.[8] These reactions to urban sprawl were common among the growing body of American environmentalists. Indeed, the historian Adam Rome has shown that the rise of environmentalism in the United States was intrinsically connected to the resistance to urban sprawl.[9] Although there is not room here to review how Gropius's later work was influenced by his environmental concern, it should be understood within this context.

An episode in 1953 may further illustrate Gropius's environmental vision. He had just completed an eight-year contract as an architectural consultant for the new Michael Reese Hospital, on the campus of the University of Chicago. When asked what would be an appropriate token of appreciation to him on the occasion of the opening of the hospital, Gropius suggested a tree bearing his name. "I want this to be a tree in which birds of many colors and shapes can sit and feel sustained," he said at the planting ceremony. "I do not wish to restrict it to species with square tail-ends or streamlined contours or international features or Bauhaus garb. In short, I wish it to be a hospital tree from which many songs should be heard, except the fake sounds of the bird imitators."[10] Environmental diversity and cultural pluralism were important goals of Gropius's visionary design.

Among the faculty at Harvard was, for a short period, Gropius's friend from his London years Serge Chermayeff, who came to view Gropius as his spiritual mentor. While teaching the introductory design course Chermayeff became deeply concerned about environmental issues related to architecture. Together with his student Christopher Alexander, he wrote a book entitled *Community and Privacy: Toward a New Architecture of Humanism* (1963), which they dedicated to Gropius.[11] They began by noting the environmental erosion of the human habitat through suburbia's invasion of farmland and wilderness. To Chermayeff this was a personal matter, as his country home on Cape Cod was in danger of being enclosed by dense development and industrialization.

Chermayeff and Alexander proposed that humans build their own autonomous ecologies instead of exploiting the natural one. There was an urgent "need to design fully functioning self-contained environments, capable of sustaining human life over long periods," rather than buildings that exploited the environment. A chief source of inspiration for such self-contained or autonomous design was research into space cabins, and Chermayeff and Alexander played an important part in transferring this know-how to architecture. "Both the nuclear submarine and the space capsule have been designed to support life over protracted periods without the possibility of escape," they pointed out.[12] Even though they saw a danger of claustrophobia in such closed environments, they believed that architects should strive toward making ecologically autonomous buildings instead of buildings that exploited natural resources. Gropius was enthusiastic about the book, which he viewed as an important contribution to the protection of the environment. "Nothing is more restoring to the inner man than great nature," he told Chermayeff.[13]

Community and Privacy was widely read, and various architects would experiment with the construction of self-contained ecological "capsules" for their clients. Patrons were not always happy about the prospect of living in an enclosed environment in order to spare the environment. It was a "difficult task of introducing the public to the notion of living in a condominium apartment in the middle of a countryside."[14] To some designers, "the ecology of privacy" suggested promising research on human behavior within enclosed buildings, for example, a student's claim to his or her own territory within a library.[15]

In the area of environmental design Ian McHarg was perhaps Gropius's most well known student. McHarg grew up near Glasgow, Scotland, where he learned to appreciate urban planning in the tradition of Patrick Geddes. From 1946 to 1950 McHarg studied architecture at Harvard, where he met with Gropius on a weekly basis in his capacity as chairman of the student council. McHarg was not an uncritical follower. During his Harvard years he questioned the lack of appreciation of vernacular design in the modernist program, and he did not refrain from criticizing Gropius on other grounds. What he did adopt from his student years was the program of uniting art and science. For example, in his research on urban planning and landscape design he studied the work of the Harvard biologist Lawrence J. Henderson.[16] While McHarg is discussed at length in chapter 7, it is important to note here that he regarded Gropius as an important source of inspiration for his later ecological design projects.

LEARNING FROM NATURE IN CHICAGO

Upon moving to the United States in 1937, Gropius secured Moholy-Nagy an appointment as the director of a new design school sponsored by the Association of Arts and Industries in Chicago. Moholy-Nagy called it the New Bauhaus. Owing to financial difficulties, the school closed down after only six months. Moholy-Nagy did not give up, however. In 1939 he started the School of Design, which evolved into the Chicago Institute of Design, which he ran until his premature death of leukemia in 1946.

Moholy-Nagy's chief patrons during these years were Walter and Elisabeth Paepcke, who through their Container Corporation of America channeled financial resources to his many projects. The Paepckes' patronage is the subject of an excellent study by the historian James S. Allen.[17] Allen shows how the Paepckes championed a more responsible and educated form of capitalist modernization through a neo-romantic interpretation of Goethe's philosophy and the German *Bildungsideal*. As Allen documents, the Paepckes were among the most important patrons of environmentalism, literature, philosophy, fine art, and music in postwar America, and Moholy-Nagy was one of their first and most important clients.

In Chicago Moholy-Nagy told his new students about the importance of environmental and social concern for Bauhaus design, and he praised Gropius for addressing these issues: "Fearlessly and uncompromisingly he defended the thesis on which the Bauhaus was built: that art and architecture which fail to serve for the betterment of our environment are socially destructive by aggravating instead of healing the ills of an inequitable social system."[18] It was important, Moholy-Nagy stressed, to think of design within the larger framework of social and environmental responsibility. An urban planner, for example, should aim to create "happy and organic cities of which inhabitants have the experience of being amidst gardens and vegetation daily, not on their weekend trips only."[19] Moholy-Nagy's famous book *Vision in Motion* (1947) was in effect, he wrote, "an attempt to add to the politico-social a *biological 'bill of rights'* asserting the interrelatedness of man's fundamental qualities, of his intellectual and emotional requirements, of his psychological well-being and his physical health."[20]

This program was spelled out in an article entitled "Why Bauhaus Education?" published in the journal *Shelter* as a manifesto for Chicago's New Bauhaus. The article was illustrated with the images of set designs from *Things to Come,* from Moholy-Nagy's London years. The Chicago students would have to read a great deal of science literature to understand the fundamental biological needs of human society,

Moholy-Nagy argued. This program was an effort to develop "a new type of engineer" who would use "an organic approach" in design.²¹ The new science of "bio-technique," Moholy-Nagy explained, "deals with transportation of natural forms and design into the media of human production. Nature evolves ingenious forms, often technologically useful. Every bush, every tree, can instruct us in and reveal new uses, potential apparatus, and technological inventions without number."²² He argued that proper teaching musters creative experiments guided by advances in engineering and science. Learning from nature was thus at the heart of the New Bauhaus program, which aimed at meeting human biological and psychological needs by combining art, science, and technology.

Evoking the Freudian formula, Moholy-Nagy modeled the teacher of architecture as a therapist breaking through the self-conscious into unconscious life forces of the student. "Once the dam of the self-consciousness is removed," he argued, "the creative energies flow with natural and remarkable ease" toward "an organic approach" in design.²³ Moholy-Nagy expressed the same ideas in the catalog for the 1938 Bauhaus show in New York by pointing to the importance of blending science, art, and craftwork. He argued that since people were "biologically equipped to experience space," a building must reflect human biology.²⁴

The journal *Shelter,* in which Moholy-Nagy published these views, was edited by Richard Buckminster Fuller, who lived and worked in Chicago in the 1930s. It is worth discussing his life in some detail, as he taught at the Chicago Institute of Design after Moholy-Nagy's death and in other ways came to represent the second generation of Bauhaus thinking about the unity of art and science by designing with nature.

BUCKMINSTER FULLER "PUTTING THE HOUSE IN ORDER"

Nineteen fourteen was a dramatic year for nineteen-year-old Richard Buckminster Fuller. He was suspended from Harvard University for irresponsible conduct and finally expelled for lack of interest in his studies. At the same time, Europe became engulfed by war, starting a chain of events that would eventually draw the failed student into military service.

Born in Maine in 1895, Fuller grew up in a family fraught with disappointments after his father's unfortunate death in 1910, which seriously hampered the family economy. His short stay at Harvard was not quite what his mother had hoped for, and upon his expulsion he found himself without connections or job opportunities. In despair he took a job at a meat-packing house in New York, and after some months he became quite desperate to join the navy.

Unfortunately, the navy was not eager to enlist Fuller, because of his poor eyesight. Still he would not give up his dream of joining the navy, and in the summer of 1915 he and his mother set out a plan to achieve his goal. They noticed that the navy needed patrol boats after the relocation of much of the fleet due to the war, and they promptly offered their pleasure boat on the condition that Fuller would be its commander. The offer was accepted, and in July 1916 he found himself in an ensign's uniform with the title of chief boatswain, restoring his boat so that it could be included in the fleet of the Navy Scout Patrol boats assigned to guard the peaceful Maine coast. On this "battleship," he later told environmentalists, he learned the military survival art of doing "more with less."[25]

One task was to rescue seaplane pilots in training, in appreciation for which some pilots would take Fuller for a flight. Later in life he often pointed to this adventurous experience as a key event. The patrol paid a small salary, life looked good, and he proposed to Anne Hewlett. They married in 1917 in a traditional military wedding, the couple passing under an arch of swords formed by his companions in arms outside the church. The wedding photo shows Fuller dressed in a stylish white navy uniform.

The dream for Fuller was to attend the temple of naval culture, the U.S. Naval Academy, in Annapolis, Maryland. The academy was keen on nurturing its officer-and-gentleman culture, and only unmarried, clear-sighted men with flawless conduct could enrolled in its four-year program. Fuller was instead conscripted into the Fourth Reserve Officer's Class, a temporary recruit school designed to provide urgently needed lower-grade officers for the war effort. In a ten-week crash course from June to September 1918 the school used the academy's campground as its duty station for training and drill. Although Fuller never attended classes at the academy, he was probably exposed to some basic lessons in naval history, warfare, and ballistics. After training he was enrolled as a lieutenant junior grade on the staff assigned to organize convoy communication under the command of Admiral Gleaves. He was dismissed when Germany capitulated in August 1919.

Later in life Fuller told audiences again and again about the paramount importance of his naval experience. He created the impression that he had commanded naval battleships, that he had received a serious education at the Naval Academy, and that he had been a commanding officer working with Admiral Gleaves. That is how he is portrayed in uncritical biographies celebrating the genius Fuller, which tell a story of noble achievements and solid education, whereas military records show the opposite.[26]

Fuller's lack of education was both a source of personal embarrassment and a motivating force behind his later thinking. He would try to hide the fact that he was almost completely self-taught by constantly pointing to the Naval Academy curricula he touched upon in the crash course in the summer of 1918. He often praised the education provided in naval training, which was aimed at solving practical, down-to-earth questions. He spoke with pride of the broad knowledge and know-how the navy allegedly had provided him: "We [in the Navy] are the only ones who can live and travel on land, on the water, and in the air," he would tell friends.[27] Most important, the war experience of collective service and sacrifice for "the human family" would reemerge throughout his life in his ideas about ecological design and management.

With the peace in Versailles, Fuller's wartime efforts came to an abrupt end. Soon he became a vagabond odd-job man drifting first to a managerial position in an armor company, next to a truck manufacturing company, and then to a job doing private airplane maintenance for a millionaire before he finally was hired by an architectural firm in Chicago, J. Monroe Howlett. There he designed, among other things, a new type of brick. The bricks failed to bring in revenue, and Fuller was laid off in 1927. In a state of deep personal crisis, he cut off his friends and was silent for a year. Fuller had acquired practical knowledge in engineering and architecture following the war, and during his year on his own he studied the needs of society and how he might contribute to meeting them based on his naval encounter and his technocratic ideas.

Fuller paid close attention to debate concerning human population growth and housing needs. One dominating view was held by followers of Thomas Malthus, who argued that the population was growing at a faster rate than food production and that humans therefore were doomed to social tragedy.[28] Fuller could not agree; following the economy theory of Joseph Schumpeter, he argued that humans could engineer themselves out of the problem. His optimistic view of human possibility was in opposition to that of the Malthusians, who Fuller believed put too much emphasis on economic and political conditions and not enough on the importance of engineering, design, and technological inventions. With regard to population dynamics, Fuller's thinking was in line with that of the Progressive Era, embodied in the thinking of Henry Ford, Diego Rivera, and Frederick W. Taylor. He shared their belief that the rationality of the scientific and managerial elite could liberate humanity from all its sufferings through machine technology.[29]

Through his newly founded 4D Company—in which he was the only employee—Fuller generated a series of innovative designs to meet the needs of a growing human population. In his view, the chief issue for designers was how to mass-produce a new type of "industrial house" built on military know-how.[30] The "4D tower house" was to prove his point. Built of aluminum, it was supposed to be a cheaper, lighter, and better building: "The largest standard one-mast model includes swimming pool at bottom deck, gymnasium, hospital, [and a] library," all of which, Fuller explained in an article in the *Chicago Evening Post*, was "to be delivered by zeppelin" to far-off locations in urgent need of reduced population pressure.[31] Following a military strategy, Fuller suggested that in order to save construction time, a massive bomb be used to blast a basin, into which the tower could then be lowered.

Judging from the "incredible smile" Fuller's drawings generated when they were displayed at the Harvard Society for Contemporary Art in 1929, his fantastic designs were not taken seriously by everyone.[32] Yet the fact that his designs were included in the exhibition is evidence of growing recognition of Fuller among avant-garde architects. It was the design for his 1927 "Dymaxion" (for *dynamic, maximum,* and *ion*) House that sparked the interest in his work. The house could be mass-produced and assembled in twenty-four hours, thanks to efficient use of materials and construction processes. Only houses made speedily on the assembly line could meet the dramatic need for shelter resulting from the predicted explosive growth of the human population. As a machine to live in, the Dymaxion House, like the automobile or the airplane, could increase the human potential, Fuller's "admirer and friend" Frank Lloyd Wright argued.[33] Others were more skeptical: "It is a man's world still," wrote a woman who thought Fuller's house, with techno-solutions modeled on military industry, had "a feminist problem."[34]

While it is true that the industrial mass production of the automobile in the Progressive Era was a major source of inspiration, it would be misleading to suggest that this alone influenced Fuller's design program. Fuller's 1932 article "Putting the House in Order" summarizes the chief aims of his design. The article starts by stating that architecture should be in the service of "Ecology," "the study of human relations, particularly as pertains to the home," a definition that resembles that of *human ecology* as used by sociologists at the University of Chicago.[35] Fuller argues that it is a matter of evolutionary survival for humankind to reorganize the social matrix away from destructive laissez-faire capitalism.

The war had taught Fuller that "with all ethical bars 'down,'" survival techniques and creative powers thrived, the public became "jazzed up" for a common good, and "science became dominantly more 'practical.'" It was from the perspective of this nostalgia for the war effort that Fuller argued that architecture should look to the military for inspiration. The wartime sense of belonging to one "Human Family" had slowly been destroyed by the culture of economic profiteering, and one should look upon architectural research as a vocation "in the service" of the public good, just as "the sailor" labored in the navy for a common cause. Fuller saw more in the navy than the spirit of self-sacrifice; he also admired the type of leadership and social order the military could offer. For him, putting the house in order meant empowering a master builder in the image of a captain. The survival of mankind from "the fright-unified Financial Fraternities strategic programs of evolution resistance" between 1918 and 1928 had been dependent on a shift in leadership, since ecological "evolution rules-out competition."[36] A combination of both cooperation and competition between individuals summarizes his view of how evolution worked.

Fuller's ecological goals included more than just helping to ease the population pressure. In the 1930s Fuller was busy inventing a series of technologies aimed at streamlining designs according to the dynamics of nature's household. The principles of nature's fluidity, energy, and aerodynamics in particular caught his imagination in projects such as the Dymaxion Car, the Dymaxion Trailer, the Dymaxion A-Frame Carrier, and even the Dymaxion Bathroom. They represented an optimistic view on the ability of design to solve the problems of population growth.

The theoretical underpinnings of these inventions are found in Fuller's book *Nine Chains of the Moon* (1938). His point of departure is again "ecology," which Fuller defines as "*the body of knowledge developed out of* the HOUSE. We stress not *housing* but the essentiality of *comprehensive research and design.*"[37] Ecology thus conceived was research on everything from human psychology, biology, and economy to material structures in alloys and how all of this was connected in new types of design. Fuller wanted to know how energy flowed through the machinery of life. His mechanistic views penetrated every aspect of his thinking, including his portrayal of humans. Borrowing from his navy vocabulary, he wrote about the soul in terms of a "Phantom Captain" commanding a ship complete with officers and sailors to maintain the body's carrying capacity and capabilities.[38] In analogy to the Phantom Captain's role of commanding the human body, he imagined architects to be the soul of the body politic. Thus, designing a house was like building a "ship" for society: humans could sail forward in the evolutionary ship steered by the Captain architect. "The goal is not 'housing,'" Fuller explains, "but the universal extension of the phantom Captain's ship into new areas of environmental control, possibly to continuity of survival without the necessity of intermittent 'abandoning ship.'"[39] Human extinction was not out of the question, Fuller argued, given environmental problems, pollution, and the general degeneration of human health resulting from the explosive population growth. Comprehensive research and design on the development of human shelters' ability to secure human evolutionary survival was the mission of ecological architecture.

Nature could guide the progressive architect, Fuller argued, since mechanical structures were the basis of everything, whether it "a book, a rose, a pencil or a baby."[40] Nature's architecture was drawn in mechanics, and the designer of technologies consequently had to look to the natural world for inspiration. Inventive (unlike destructive) industrialization was both "organic and evolutionary," Fuller explained in an article in *Fortune,* since it transformed the human species "far above the levels of the past" and thus could lift the country out of economic depression.[41] Studies of movement and the transformation of energy were particularly important for understanding this process, since energy was the vehicle for evolution toward a more loving society. (Fuller defined love as "radiation of pure energy," energy he sought to mobilize through his design.)[42] His broader aim was an ecological "Streamlining of Society" so that energy would transfer smoothly.

Most of the Dymaxion technologies were commercial failures; it was not until the development of his Dymaxion Deployment Unit, or DDU, that Fuller realized his dream of seeing houses roll off the production line. With Europe becoming involved in another war, the United States soon became a prime provider of materials for the Allied forces, and the DDUs met a pressing need for easy-to-assemble shelters. The most ingenious aspect of the shelter in terms of ecology was its natural air conditioning; Fuller used the sun to create air pressure that in effect cooled the building. Hundreds of these homes were mass-produced and used as radar shacks and for other military purposes. Fuller soon made sketches of a civil version of the house to be mass-produced at the airplane factories in Wichita, Kansas, where workers were in urgent need of housing. Collaborating with the factory, Fuller made a prototype of a new Dymaxion Dwelling Machine, which went on display in the fall of 1945. This so-called Wichita House, paid for by the air force, was never put into mass production, yet Fuller's inventive ideas propelled him into teaching positions first at the Black Mountain College, in North Carolina, in 1948 and subsequently at Moholy-Nagy's Institute of Design in Chicago in 1949. In Chicago he would teach according to the syllabus of the late Bauhaus designer.

The Dymaxion inventions were all examples of technological solutions to global and local ecological problems as Fuller saw them in the 1930s and 1940s. They emerged from the navy's ethos of sacrifice and service for the common good, and the DDU's initial success was secured by military needs. As machines for living the Dymaxion inventions represented an optimistic view of the ability of technology to solve the growing ecological problem of population growth.

HERBERT BAYER'S GLOBAL DESIGN

Herbert Bayer's solutions to ecological problems differed from Fuller's technologically informed designs. A former faculty member of the Bauhaus school's printing department in Dessau, Bayer would instead emphasize the importance of visual communication in changing people's way of thinking about nature.

According to his first biography (published in 1947), Bayer was born in 1900 in a village near Salzburg, where he "grew up in the atmosphere of the Austrian Youth Movement, which was a typical outgrowth of [the] Romantic search for freedom from an inherited mode of life. The 'new man' was supposed to reach a purer state of inner harmony and vigorous independence by developing his own creative powers."[43] Although such lofty language may sound strange to our contemporary ears, this was definitely the way Bayer wanted people to view his background. He saw himself as someone who early on had embraced the romantic call for a new harmonious beginning for the individual's relationship with the social and the natural world. It is necessary to review this call in some detail to fully understand his designs with nature and his environmental cartography of the 1950s.

The horrors of World War I taught the young Bayer that national and cultural isolation had to yield to global understandings of human relationships. As the Austro-Hungarian Empire fell apart, he followed the universalistic thinking of the legendary avant-garde Austrian coffeehouse discussions.[44] These debates were dominated by the Vienna Circle's logical positivism and the eye-opening psychology of Sigmund Freud. Bayer was in the midst of this creative cultural upheaval, working first as an apprentice in graphic design and then in an architectural office from 1919 to 1921. It was during this period that he came upon a flyer titled "Bauhaus Manifest," written by the architect Walter Gropius to promote his school. Located in Goethe's Weimar, the school called for a revival of the rich relationship between humans and nature that the philosopher had promoted. This was to be achieved through a unity of arts and crafts at workshops that sought to unify the decorative arts with a universal industrialism.[45] Bayer arranged for an interview and was accepted by Gropius as a student in the four-year program.

Bayer believed that the full attainment of the human potential was to come in designs based on a union between the sciences and craftwork, and he laid special emphasis on the insights of Freudian psychology. He became a student of the artist Wassily Kandinsky, whose constructivist abstractions Bayer sought to apply to social problems and realities.⁴⁶ He was also stimulated by the bionic approach to design taught by Moholy-Nagy, who was inspired by the biological sciences to generate functional forms.

The goal of creating a universal and objective science-based design would dominate Bayer's work after his final examinations in 1925. He accepted the position of director of the new printing and advertising workshop at the Bauhaus's new location in Dessau. There Bayer designed a new typography based on the "geometric foundation of each letter," which was meant to liberate the human mind from the burden of traditional ornamental typography.⁴⁷ To make the writing experience even more efficient, he abandoned capital letters altogether. Thus, Bayer would type:

"bauhaus gave me a way of life."⁴⁸

He called it the "universal type," since it was based on geometry, as if to reflect the positivist attempt to generate a universal scientific language of logic. The typography was adopted by most Bauhaus publications, and it soon became the trademark of the school.

Bayer taught at the Bauhaus until 1928, after which he became the director of *Vogue* magazine in Berlin, as well as editor of the influential avant-garde design journal *Die neue Linie*. In the latter position, Bayer enjoyed the company of the surrealist romantic Max Ernst, who sought to bring out the savage within by arranging "Walpurgies Nacht orgies with nude girls jumping over fires."⁴⁹

The 1938 Bauhaus retrospective at the Museum of Modern Art in New York, which Bayer designed, gave him the opportunity to get out of Nazi Germany. In the show he tried to show that design could be a tool for making the world better by mobilizing the physical, rational, and emotional aspects of the human condition. For example, he arranged the layout of the exhibition according to human biology; it was designed so that the viewer would move from left to right, just as he would read from left to right, so that mind and body would function together. He also introduced images and objects in line with Freudian ideas about human perception. The conscious and the subconscious were to be mobilized by displaying images on the floor, walls, and ceiling, a method he borrowed from the artist El Lissitzky.⁵⁰ Bayer believed that the conscious and subconscious experiences of the viewer would be enriched by iconography in which one form dominated, with smaller illustrative forms contributing psychological and compositional animation. In this way he tried to engage viewers physically, emotionally, and rationally and thus achieve the Bauhaus goal of designing for a "complete being."

In New York Bayer tried "to overcome the traditional forms of pictorial presentation" by illustrating events through "a functional vision," according to the original meaning of Louis Sullivan's motto "Form follows function."[51] The standard Bauhaus interpretation of Sullivan's motto followed Moholy-Nagy's reasoning that functionalism should be understood in terms of phenomena occurring in nature, where every form emerges from its proper function. Graphic design would be functional, Bayer believed, if its form followed human conscious and subconscious reactions to light and structure. In his design he would strive toward a simplified graphic environment that could improve human functioning in a dramatically changing social and natural world.

Bayer's Nazi experience and his arrival in New York gave him a new sense of universal responsibility. A better world was possible in an environment that embraced all human abilities. Bayer designed the book cover to the architect José Luis Sert's famous book *Can Our Cities Survive?* (1942), for example, because it addressed the problem of slums and overcrowding and saw urban planning and modernist architecture as the remedy.[52]

Bayer promoted these ideas about functionalism through "activities in a hundred-and-one fields" of design while living in New York, including the exhibitions "Arts in Therapy" (1941) and "The Road to Victory" (1942) for the Museum of Modern Art. It was Bayer's ability to create a dynamic and integrated viewing experience that made these shows a public success.[53] The culmination came in 1943 with "The Airways to Peace" exhibition, whose focus on engaging the viewer's mind, body, reason, and emotion in his or her experience of another world war illustrated Bayer's thinking about the relationship between humans and the global environment.

"Global war teaches global cartography," claimed an article in *Life* magazine. The article came to the attention of Monroe Wheeler, the director of the Museum of Modern Art, who decided to produce a show about it.[54] It was one of many popular articles appearing at the time in

Life, where Fuller worked as a journalist, and that claimed that the a new sense of global understanding of the world emerged with the growing use of the airplane.[55] This was also the opinion of the Consolidated Aircraft Corporation, which in advertisements claimed that "No Spot on Earth is More Than 60 Hours From Your Local Airport." In a similar vein American Airlines printed advertisements illustrating the airspace with the text "We exist *upon* one globe, and *inside* another globe."[56]

As the patrons for "The Airways to Peace" exhibition, these companies pushed for a show that associated globalization with a peaceful future. The "air age geography" was to bring a new age of intercontinental understanding between peoples of the earth.[57] Wheeler's plan was to present images of the Allied forces in the different theaters of operation around the world, followed by a geographical and ecological explanation of how airplanes and the war made the world one community. The aim was to show the audience that thanks to the airplane, people were no more than a few hours apart. This idea of "a show of world geography and ecology has elicited . . . enthusiastic endorsement in all quarters," Wheeler told the Office of War Information.[58]

In order to link the ecology of the earth with the movement of airplanes in airspace, Bayer, who designed the show, turned to the meteorological sciences, as airplanes were dependent on weather. From then on, meteorological representation would be a major element in his work. The show featured posters indicating the movement of wind and airplanes, as well as the wind's role in various environments on earth and in different battle locations. The movements of wind and planes above land on which soldiers fought were represented in contrasting colors on maps or diagrams.[59] This intermix of weather, geography, airplanes, and warfare constituted Bayer's understanding of the global ecology.

The central element of the show was a floor-to-ceiling globe in which the audience could stand and thus get a panoramic view of the world as a whole. Using the psychological technique of total perspective developed in the Bauhaus retrospective of 1938, the globe invited the audience to experience the world as one entity in which they were at the center.[60] This experience was reinforced by the catalog text, which stated that humans were moving toward a global understanding of the world: peace would come through a global organization such as the United Nations, imperialism at home and abroad would end, and all peoples of the world would strive together for the common good.[61] Bayer's macrocosmic design also inspired models for understanding the microcosm of cells, such as the walk-through museum model of a cell made by the famous science artist Will Burtin.[62]

The romantics' call for mobilizing the complete human being as this was understood in Bauhaus design was the context for Bayer's construction of a global world in the unusually popular "Airways to Peace" exhibition, which traveled to major cities in North America. According to Wheeler, it embodied "the global concept."[63] It became Bayer's point of departure for his graphic, architectural, and artistic expressions of a harmonious relationship between humans and the natural world.

4. THE GRAPHIC ENVIRONMENT OF HERBERT BAYER

It was Herbert Bayer who introduced modernist imagery to illustrate conservation values and environmental cartography in the United States. His graphic work represented a neo-romantic attempt to reconcile capitalism with humanistic values and the conservation of environmental resources. By appealing to the intellectual, physical, and emotional sides of the observer, Bayer sought to harmonize the humanist legacy with industrialization of the natural world. Measured in terms of dissemination, his graphic design became so widespread in environmental debates that few today question the origin of this style. The widely used recycle symbol may serve as an introductory example.

The two dominant areas of Bayer's work were designs with nature and cartography. His designs with nature were based on a Bauhaus vision of a new kind of industrial humanism that entailed a life in harmony with the social and the natural world. An example is his famous *Grass Mound* (1955), which came to inspire a whole generation of earthworks artists, who literally broke the ground for the ecological-design and restoration projects of today.[1] Bayer's designs with nature were an integral part of a neo-romantic program of humanism and environmentalism promoted by his chief patron, the Container Corporation of America.

Bayer's impact is perhaps most apparent in the field of cartography, however. In his *World Geo-Graphic Atlas*, published in 1953, Bayer established a Bauhaus iconography in environmental atlases. While for historians of graphic design the *World Geo-Graphic Atlas* represents "an important milestone in the visual presentation of data," it has not received attention from historians or sociologists of cartography.[2] The atlas has only been used as evidence for what it claimed to be, namely, a collection of facts. Historians of science and cartography have exposed the rich layers of social power that maps embody.[3] Bayer's graphic environment came to claim the land. The artist has the ability to influence how people see and interact with nature, and Bayer's vision of the world as a whole entailed a humanist approach to design with nature based on a neo-romantic modernization of environmental cartography.

DESIGNING HARMONY IN ASPEN

After his move to New York in 1938 Bayer did not feel that he was making progress toward his ideal of a "complete human being." That full life came through an opportunity to work as a designer in Aspen, Colorado. There he was able to focus his global perspective on local agendas, such as the recycling of natural resources and ecological design.

The patron who made the Aspen experience possible was Walter P. Paepcke, the chief owner and director of the highly successful Chicago-based Container Corporation of America, known for introducing cardboard boxes to the United States. He was an unusual business leader who in a spurt of postwar enthusiasm came up with the idea of building a nature and cultural resort in Aspen. Using his company fortune, Paepcke bought a significant portion of the town of Aspen, while his wife supported him with her social abilities and taste for the avant-garde. Together they tried to heal the wounds of World War II by organizing various conferences and cultural festivals in Aspen aimed at educating the social and financial elite in the virtues of democratic values, love for nature, and philosophizing in what evolved into the Aspen Institute for Humanistic Studies.

The Container Corporation wanted to be associated with high-quality design as well as social and environmental responsibility, which Paepcke believed could help "to 'break the ice' when our salesman calls on his prospect."[4] The program was managed by the chromatologist Egbert Jacobsen, in the company's department of design. They engaged a host of avant-garde artists to develop the corporation's trademark, including (besides Bayer) Will Burtin, Miro Carreño, David Hill, William de Kooning, Fernand Léger, Henry Moore, and Man Ray. They all created advertisements reflecting social or artistic topics of their choice. As a result, the company was hailed as having "the most creative program in today's advertising," thanks to its use of Bauhaus designs, which Paepcke nurtured as the principal patron of Moholy-Nagy's New Bauhaus school.[5]

The famous Swiss art critic Sigfried Giedion viewed with suspicion this willingness to serve commercialism and published a damaging review of Bayer.[6] Bayer saw it differently. He believed that an artist could most effectively engage society at large on important topics by working for a commercial company. One of his chief concerns was recycling and resource management. With this in mind he designed a series of eleven advertisements for the Container Corporation, nearly all of which focused on the importance of recycling. Here he followed Paepcke, who in the 1930s had made the strategic mistake of underestimating the importance of owning vast timberlands to support his pulp mills. He tried to remedy the problem by producing cardboard from wastepaper, and the result was a highly successful recycling program. By 1941 the Container Corporation produced 90–95 percent of its cardboard from wastepaper, which, as a *Fortune* article noted, meant that the company was living "over and over again upon its own waste." It was also "in the tick of the war business," producing cardboard boxes for everything from boots to bombs.[7] In support of the nationwide war effort, Paepcke in a series of advertisements pushed further for collecting wastepaper, and Bayer's work supported the agenda. One of his advertisements reads, "paper that goes to war is paper that wasn't burned. Save waste paper! sell or give to local collectors." The text is illustrated by objects arranged in an elegant S shape that show the process of turning wastepaper into cardboard boxes for bombs dropping from a plane (Fig. 1).

Judging from public reactions to them, Bayer's advertisements for recycling were a success. In the spring of 1945 Paepcke decided to sponsor an exhibition at the Art Institute of Chicago (which worked in association with Moholy-Nagy's School of Design, which Paepcke sponsored) of the entire wartime advertisement campaign, and he hired Bayer to do the job.[8] The result was "Modern Art in Advertising," a traveling exhibition that had a record-breaking two hundred thousand visitors.

During the summer of 1945 Papecke pressed Bayer to move to Aspen and be the town's architect, designer, and resident artist. His job would be to help transform "the old ghost town of Aspen in its material, social, and communal aspects."[9] Since the heyday of mining, Aspen's population had declined steadily from seventeen thousand inhabitants to only eight hundred. Paepcke saw the town, with its natural beauty, as a future vacation spot. By autumn a ski slope was under construction. Bayer had been contemplating leaving New York to move either back to Austria or to Mexico. Aspen, Paepcke argued, had "the best skiing conditions," and moving there would be almost like moving back to his Tyrolean homeland.[10] As Bayer continued to consider where to move, he fell seriously ill from exhaustion, and his friends, including Gropius, wondered whether he would survive another year of Manhattan working hours. His wife, Joella, came to see the American Rockies as his remedy, to which Paepcke responded: "We will all have to gather around him and shout: 'Go west, young man!'"[11]

Aspen became a place where Bayer could in effect nurture his ideal of living "in the most human way," which meant hiking, skiing, making art, writing, architectural work, graphic-design jobs, family life, and enjoying good wine.[12] His duties for the Container Corporation and Paepcke's Aspen Company included everything from town planning and architecture to designing stationery for Paepcke's hotel. Bayer found his job to be "spiritually as well as physically" lucrative, and his patron soon became a close friend.[13] Intellectual discussions at the Aspen Institute and attending the Aspen Music Festival were part of Bayer's new agenda. Among the numerous artists he met were the nature photographer Ansel Adams, with whom he discussed techniques for capturing the sublime wilderness on film.[14]

Bayer's decision "to seek his artistic sustenance in nature" was, according to Egbert Jacobsen, vital to understanding Bayer's work in Aspen.[15] "The participation in shaping an environment, in dealing with social problems [and] in building a community life" were Bayer's motivating forces.[16] As a token of his goodwill toward the citizens of Aspen, Bayer offered to redesign houses and offices free of charge, to which the *Aspen Times* responded that it was "exceptionally good luck to have one of the world's great designers in our midst."[17] From Paepcke's perspective, Bayer was living proof that Aspen had more to offer than unemployed miners.[18] Bayer's activities bore fruit, and Aspen "came back" to life, Bauhaus style.[19] In his artwork he drew inspiration from the natural sciences, especially geography and meteorology, to produce effective abstract landscape art, most famously his sgraffito mural in the Seminar Building at the Aspen Institute for Humanistic Studies.

As an architect Bayer tried to design buildings that "extend into the natural ecology."[20] At the Aspen Meadows Campus, the Institute for Humanistic Studies and the Aspen Art Institute were built according to Bauhaus principles, resulting in designs users considered natural and thus ecological.[21] To design in harmony with nature was in agreement with Bauhaus ideals. Both Moholy-Nagy and Gropius had expressed similar concerns; however, later attempts to design with nature were not in line with their views about ecological design. Unlike ecologically minded architects and artists of the 1970s, Bayer pursued a strictly anthropocentric approach. "In respecting nature," he argued, "the artist will not imitate nature but create a spiritual world of itself side-by-side with nature, . . . [since] both natural environment and man-made environment can exist with each other if their boundaries are understood."[22]

Bayer's *Grass Mound*, built at the Aspen Institute in 1955, may serve as an example of the kind of spiritual world Bayer imagined. Placed in a scenic environment, it consists of a grass mound measuring forty feet in diameter with a heap of soil, a pit, and a rock of raw marble in its midst. It is a human space in nature upon which to reflect on the potentials of artistic agency. Agency was to Bayer a matter of mobilizing the full human potential, as viewed from the Freudian perspective. Experiences of agency in nature should be viewed accordingly, and not in terms of imagined or real natural powers. A manipulated photograph from 1959 showing birches with gazing eyes, for example, can be understood as a depiction of the psychological condition of paranoia, with agency sensed in every tree of the forest. The humanism of Bayer saw biocentric notions of agency as poor epistemological or psychological understandings of boundaries between humans and the environment.[23] The belief that humans could only see the world from a human perspective was at the core of his cartography.

A NEW GRAPHIC REPRESENTATION OF THE WORLD

"Geography is man-made stuff," Bayer would say, quoting the American geographer George Renner, "and therefore its basis must be resurveyed and re-evaluated over and over as times and the instruments of power change."[24] As a mapmaker designing a new graphic representation of the world, Bayer was well aware of the power of inclusion, exclusion, perspective, and emphasis. He therefore made explicit his artistic point of departure by using the hyphenated form *geo-graphic* in the title of his *World Geo-Graphic Atlas*, of 1953, since the science of *geography* was to be based on a *graphic* footing.

To illustrate his responsibilities in the atlas, Bayer wrote that "according to ancient myth, Atlas was a Titan condemned to support the heavens on his shoulders," in a reference to the power of representing the world with graphic images.[25] This statement evoked much wit in Aspen circles, as Paepcke would tease: "I didn't realize that Herbert thought he was Atlas carrying the world. I just thought that he was Herbert carrying the World Atlas."[26]

The atlas was Bayer's most ambitious project ever, with a significant cost, even for a wealthy client whose "complete trust" he enjoyed.[27] It was to be produced privately as a gift for customers of the Container Corporation, reflecting the fact that boxes tend to move over long distances. He began in 1947, and it took five years, during which time he hardly had any other design activities. He was assisted by three designers, Martin Rosenzweig (1947–49, 1952–53), Henry Gardiner (1949–53), and Masato Nakagawa (1952–53), as well as a secretary, a proofreader, and a copywriter. More time and money went into this atlas than into any other atlas of the period. The production studio in Aspen was like a research laboratory, aimed at making an atlas of the highest professional quality. The result was a 368-page book that included 120 full-page maps, 1,200 smaller maps, and 4,000–5,000 finished drawings (including separate ones for images with several colors).

In a lecture entitled "Goethe and the Contemporary Artist," delivered in 1949, Bayer sought to make a case for his humanistic view of nature by letting art take the lead in the production of scientific knowledge. A visual language for the geological sciences had originated with the romantics, and it was therefore no accident that Bayer evoked Goethe's authority.[28] The agenda of the atlas was to follow the advice of the philosopher, namely, to nurture a fully integrated human life in harmony with the natural world.

Harmony was to be restored through the conservation of energy and material resources, and the atlas's graphic design was to incite readers to protect the environment. Bayer pointed to "unmistakable signs that the climate of the North Atlantic region is growing warmer" owing to "the progressing depletion of its [American] resource base." In the case of Germany, "lack of essential raw materials and lack of 'lebensraum' for a growing population led to disastrous attempts to secure these needs."[29] "Destruction of resources is as old as mankind," Bayer wrote, "but it is the special characteristic of the 19th and 20th centuries: no problem confronting the world today is more vital than conservation and wise utilization of natural wealth."[30] He followed a neo-Malthusian line of argumentation: the dramatic population growth made solving the problem essential. Bayer devoted the final pages of the atlas to a call for action. Among the topics broached were the following global environmental problems facing humanity: limited availability of land, the abuse of forests, the limited reserves of minerals, soil erosion, and the widespread abuse of energy.

The atlas was to provide an artistic form of therapy that might facilitate solutions to the environmental crisis. The dominance of textual over imagerial reasoning had made humans one-sided, Bayer argued, resulting in a lack of appreciation for nature. What was needed was a return to the primitive appreciation of images. This would bring forth a more balanced human being and consequently a society that treated nature with respect. Bayer adapted an evolutionary view of language by arguing for a liberation and enrichment of the human potential through graphic design. Over thousands of years, he argued, humans had become "letter-poisoned" by textual communication. Sound and visual signs made up the original human language, and one should therefore appeal to the savage within.[31] By evoking this romantic longing for primitive sign language, the atlas aimed at a nobler mode of human communication. The images were to concentrate the message and liberate the reader from the burden of textual information. Simple images had the potential of bringing out that Edenic human language that had been blurred by the Babylonian confusion of tongues. He envisioned, for example, an improvement in human relations if businesses could communicate their messages through subtle trademarks instead of an overwhelming flow of textual information. In creating the atlas, Bayer drew on his experience in designing exhibitions, such as the "Airways to Peace" exhibition, in "the activation of the white areas, the principle of contrast as vitalizing element, the idea of visual continuity through the pages, the use of pictures, the influence of montage by fusing various elements into superimposed images, change of scale within the type faces, and so on."[32]

To facilitate this transition to a language of images Bayer developed a set of environmental symbols. They were meant "to tell the story in the simplest terms" so that the reader would get an "immediate comprehension" through just a quick glance.[33] In creating the symbols he would first try to strip a given problem down to the essentials by forming the symbol as a functional representation. In the case of metals, for example, Bayer used an ingot with abbreviations from the Periodical Table. With respect to aqua- and agricultural foodstuffs, as well as various industrial and commercial products, he also tried to create symbols that could be understood by illiterates. They were placed on the maps to indicate an important activity for a certain place, while rows of symbols beside a map represented related statistical information. In this way he made statistical information more easily available by replacing text and numbers with graphic symbols. On the last page of the atlas is a graphic illustration of future world populations in which Bayer used the image of a person to represent 100 million people and represented exponential growth over time with additional bodies and a dramatic, thickening red arrow (Fig. 2).

Bayer used the human body as an image of population throughout the atlas, and the final illustration is a graphic summary of the problem of population growth with respect to resource conservation. Both the symbol and the arrow were copied in later discussions of population growth, and Bayer's use of human bodies became a sort of trademark of overpopulation.

The symbols and images in the atlas were meant to represent minor and major environmental histories. As Bayer explained, "It was the story in the image which we looked for, not the image itself."[34] In this dynamic (as opposed to static) view of design Bayer was following the artistic technique of Moholy-Nagy. In his photographic art, it is worth recalling, Moholy-Nagy tried to capture the forces of evolution in action. Similarly, Bayer saw modern art, through "the study of the living shape," as an expression of the dynamic forces in nature.[35] In the atlas these forces were expressed in the narrative of the earth's origin and possible end. Bayer took the reader through the earth's beginning as a cloud of dust and its continuing astronomical, geological, atmospheric, and evolutionary history, ending with discussions concerning the need for conservation of the world's resources in view of the dramatic population growth. The shorter narratives within the atlas were to support this view through information about movements of goods and people from one region, land, or continent to another in order to convey a story of a dynamic earth in constant social and natural evolution.

The colors of the atlas were based on a harmonized universal system of charts with precursors as far back as Goethe's chromatology. Bayer followed the Container Corporation's color policy, developed by his colleague Egbert Jacobsen. In cooperation with Gropius, Jacobsen had worked out a system for the company published as *The Color Harmony Manual* in 1942, with enlarged editions in 1946 and 1948. By 1953 more than two thousand advertisers, printers, publishers, architects, artists, designers, industrialists, mechanists, paint manufacturers and dealers, schools, and textile producers owned the manual. It thus had a significant effect on the use of colors in postwar America.[36] At the Graduate School of Design at Harvard, for example, the manual was praised as a "much treasured and much used" tool.[37]

Jacobsen sought a scientific basis for color analysis based on the chemist Friedrich Wilhelm Ostwald's hue circle. "We need no longer wander in a chaos of conflicting color impressions composed of rainbows, Christmas ties, and ink swatches," Jacobsen argued. "We now have an orderly concept which enables us to understand color relationships and, therefore, eventually to combine colors with some hope of producing harmony."[38] Jacobsen's notion of harmony was a sense of order and completeness induced by an experience of colors. He argued that a confusing use of colors often reflected a deeper sense of social discontent and that by an orderly use of colors design and aesthetic expression could contribute to the betterment of the world. The color system also bettered the Container Corporation's image: it distributed *The Color Harmony Manual* to its clients for free, while those who were not within its core network had to pay $125 for it, so that the manual was seen as a real gift, not just company advertising.

Bayer followed *The Color Harmony Manual* on every page of the atlas, thus making sure that the colors used to represent the environment would be in harmony. Red, blue, green, yellow, and so on, were used so as to have the best psychological effect on the reader. When Bayer presented the atlas in his lectures, he emphasized that he had considered both the "psychological properties *and* the esthetics of color harmony" in determining which colors to use.[39] Because of the psychological sensation they created of advancing, owing to their long rays of light, warm reddish hues were used to illustrate movement or high altitudes. By contrast, because cold bluish hues, owing to their short rays of light, created a sensation of retreating, they were used to illustrate immobility, low altitudes, or water. Bayer would contrast these colors according to *The Color Harmony Manual* in order to construe a psychological sense of balance and order in the natural world, using green, yellow, and brown as in-between colors. The result was a design differentiating the physical attributes of nature through the use of colors with increasing or decreasing intensity following geographical contour lines. "At a single glance one can see where mountains are highest and the sea deepest," Moholy-Nagy's wife Sibyl noted in her praise of the atlas.[40]

The illustration entitled "Overseas Emigration from Europe (1820–1937)," in blue, green, and red, may serve as an example of how his use of colors also could tell a dramatic history of movement of populations to Australia, South Africa, Latin America, and North America (Fig. 3). The color coding represented the bottom (blue), middle (green), and top (red) of the world, while the thickness of the arrows indicated the number of emigrants from the "severely overpopulated" European nations.[41] The illustration was one of many addressing population dynamics and their dramatic environmental impact.

In the atlas different realms of knowledge were "to be fused into a coherent entity," thus creating interdisciplinary understandings and perspectives of the world. In this way the implementation of the Bauhaus ideal of designing for a complete human being was to bring forth an integrated view of the globe. "Swiftly spreading global communication and the increasing interdependence of all peoples compel us more than ever to consider the world as one," Bayer argued.[42] This social globalization was also relevant for adapting an integrated view of the sciences. The front page of the atlas has an image meant to capture the integration of astronomy, demography, geology, geography, economics, and climatology as "A Composite of Man's Environment" (the subtitle of the book). Each realm of knowledge is represented by a circle with overlapping colors, and a human being is located at the very center of the image (Fig. 4). The human centeredness of the atlas was deliberate, as Bayer saw human agency at the heart of both scientific and artistic practice.

Bayer was untrained in reading scientific texts, and it was thus a challenge for him to determine relevant data in various fields. "I felt heavy responsibilities all through the process of making the book," Bayer later confessed. He initially asked scientists of various disciplines to contribute to the volume but discovered quickly that the information he received was useless, irrelevant, or at odds with his own vision of the world. He therefore did most of the research himself, a process he described as "a good adult education." The result was a synopsis of what he as a designer thought to be the most relevant information for readers. "A scientist would not think in terms in which I worked," he argued, since scientists tended to publish their research in "unimaginative textbooks, specialized papers and journals."[43]

Bayer put his ideal of living an integrated human life into practice by determining the relevance of scientific evidence himself. He used a host of scientific sources, mostly from geographers, whose work he used rather selectively. For example, he evoked the work of Ellsworth Huntington, turning the geographer's environmental determinism into a possibilist perspective. Huntington argued that the distribution of human health and energy on the basis of climate could explain the social levels of civilizations, measured by the number of inventions, the power to lead, and, above all, the trading of goods and knowledge.[44] Bayer used Huntington's framework of analysis but emphasized that it was energy produced through human agency (not climate) that determined the fate of civilizations. Energy produced by humans was the chief source of energy for various industries, which could harm the environment in various ways. North Americans, for example, were about to become "energy slaves" of their power-hungry machines and were in urgent need of more energy-efficient technologies.[45]

The *World Geo-Graphic Atlas* was published in 1953 in an edition of thirty thousand copies, which were distributed exclusively by the Container Corporation. A change in environmental policy would come from the industrial and political elite, Paepcke believed, and not from ordinary customers in bookstores. Indeed, he dismissed several offers from commercial publishers seeking to bring out a trade edition of the atlas.[46] The elite included, among a host of dignitaries, the 1952 Democratic presidential candidate, Adlai Stevenson, who was a personal friend of Paepcke's.[47] Most of the copies, however, were sent out as company gifts to the corporation's customers. The Walter P. Paepcke Papers include a significant volume of letters expressing customers' excitement and gratitude. It also contains numerous letters from people explaining why they too should receive a copy. An investment officer at Yale University did not receive a copy as a wedding gift, for example, nor did the dishwasher at the Jerome Hotel in Aspen who in a moving letter pleaded for a copy for his daughter. A graduate student about to go to India to teach geography was also rejected. On the other hand, a representative at the Great Book Foundation, a professor of geography at the University of Hawaii, and the American ambassador in Teheran all got complimentary copies. In some cases subtle and not so subtle gift exchanges were involved, including hospitality, artwork, and introduction to VIPs. It was all done in accordance with Paepcke's belief that the romantic modernization of human relationships with the environment would have to start with the industrial and social elite.

The atlas was well received not only by recipients but also by reviewers, according to whom it was "surely one of the most edifying and beautiful books ever printed," with images that "tell more in pictures than in words" the important "facts of conservation" and "the vital statistics of every man's essential needs."[48] In scholarly journals the atlas was judged to be "the handsomest and best atlas ever published in America," though Bayer was criticized for including too much "peculiar information."[49] Designers, on the other hand, hailed the atlas for its "direct visual communication" as "what a 20th-century atlas should be," with maps surpassing "any ever shown in an American atlas to a degree which is almost embarrassing; they are masterworks of the cartographer's art."[50] A Swiss designer saw the atlas as an example of how Americans communicated, "simply, directly, and with all possible forcefulness," in contrast to "the pedantry and conservatism of the Old World."[51] Moholy-Nagy's wife, Sibyl, saw the atlas as part of a larger "powerful trend toward visualization" at the expense of the old-style authority of texts. The result was "the first integrated world picture" of the environment as a whole, she claimed.[52]

The *World Geo-Graphic Atlas* had a lasting impact on environmental cartography. Atlases of world resources produced before Bayer's publication had made little use of modernist graphic language.[53] This would change with environmentally informed atlases of the 1970s, which borrowed extensively from Bayer in their integration of color, graphics, and symbols.[54] For example, Rand McNally's bestselling *Earth and Man World Atlas,* published in 1972, basically copied Bayer's graphic collage technique.[55] Juxtaposing images and maps became a way for cartographers to illustrate their environmental concerns in maps.[56] The six editions of Ben Crow and Alan Thomas's *Third World Atlas* (1983–88), produced by Ros Porter, owed much of their success to the graphic design of Bayer's atlas.[57] The technique was also used in the widely read *Gaia* atlases and the *New State of the World* atlases, which in the 1980s and 1990s came to dominate the market for maps depicting the environmental crisis.[58] Yet unlike Bayer, the designers of these atlases did not have much faith in the industrial elite. As the atlas historian Jeremy Black has pointed out, their graphic symbols were "employed to drive home points" about business culture and companies conducting "organized crime" against the environment and the poor.[59] Although Bayer introduced the graphic methodology for environmentalist cartography, he was largely ignored by a generation of mapmakers who hoped for a revolution from below.

THE GRAPHIC EMPOWERMENT OF ENVIRONMENTALISM

The adoption of Bayer's graphic environmental language indicates that environmental debates are more indebted to artistic communication than their followers have been willing to admit. His environmental humanism empowered a series of environmentalists, including Elisabeth Paepcke, who, after her husband's death in 1960, became a principal patron of the Thorne Ecological Foundation (from 1966), the Seminar on Environmental Arts and Sciences (from 1967), the Aspen Center for Environmental Studies (from 1968), and the Aspen Institute for Humanistic Studies. Through these institutions the relevance of humanism to the environmental debate became apparent for a new generation of ecological thinkers, including Oakleigh Thorne, Frank B. Golley, John McHale, and Donald Worster. This patronage in itself constitutes a rich history of resistance against developing Aspen into the jet-set resort it has become today, and it deserves its own analysis.

Bayer's global humanism and environmental designs created a visual language of colors, images, symbols, and dynamic illustrations with the goal of a harmonic relationship between humans and the natural world. The full import of his work has yet to be understood, though it is safe to say that he had a significant impact on environmental cartography. That he inspired others working in the area of visual communication of environmental concerns is evident in the case of the symbol for recycling. In celebration of the first Earth Day, in 1970, the Container Corporation announced a design competition for a trademark for recycling in the spirit of Bayer. The competition was won by Garry Anderson, a student at the University of Southern California, who presented the symbol at the Design Conference in Aspen.[60] The Container Corporation did not copyright the trademark but put it in the public domain so that everyone could use it free of charge. Based on the Bauhaus ideal of living in harmony with the natural world, it is now universally known, as it is printed on millions of recyclable products all over the world.

5   BUCKMINSTER FULLER AS CAPTAIN OF SPACESHIP EARTH

"We'll be remembered as those who lived in the age of Buckminster Fuller," the American poet and composer John Cage noted in his renowned "Diary: How to Improve the World (You will Only make Matters Worse)," of 1971.[1] There is some truth to Cage's observation, as Fuller's various technological designs for a better environment did capture the imagination of the counterculture of the 1970s. But did his answers to "how to improve the world," in Cage's words, "only make matters worse"?

Architects, designers, and historians have argued that Fuller did improve the world. In the current literature he is portrayed as providing designers and environmentalists alike with viable alternative tools to improve life on Earth. One historical study, for example, argues that his Dymaxion maps and World Game "conveyed a strong message about the value of democratic participation" and that his famous domes "expressed community, gathering and sharing" as a metaphor for spiritual ecological "earth-consciousness."[2] In a similar strain of reasoning, the architect Norman Foster points to Fuller as a source of inspiration for his own dome design for the newly renovated Reichstag in Berlin. Indeed, he codesigned an "autonomous house" with Fuller in 1982.[3] A host of websites, republications, video productions, television programs, anthologies, and hailing biographies of the last decade provide further evidence of Fuller having a significant contemporary following.[4]

I question this admiration of Fuller in view of his authoritarian program for saving the earth from ecological disaster. As discussed in chapter 3, Fuller developed an early admiration for the naval style of leadership during World War I. This admiration was reinforced by his subsequent interest in ideas for scientific and technocratic management in the 1930s. During World War II and the cold war, Fuller emerged as a major designer of military tools. In the 1960s he fashioned himself as a naval captain, designing energy-saving buildings and conducting military war games to save Spaceship Earth from ecological destruction. Fuller's "high-modernist schemes" for improving the world, to borrow James C. Scott's terminology, inspired environmentalists in 1970s.[5] Contrary to the widely held belief that this social movement represented a historical break with the past, followers of Fuller represented more of a continuation with previous high modernism. Fuller came to fashion his environmental supporters along the lines of a naval community based on the military values and the managerial sciences of the 1930s. Ultimately Fuller's eco-utopia was to be a world without politics; at the helm would be designers and ecologists guided by the captain of Spaceship Earth.

FULLER AS A COLD WAR DESIGNER
FOR THE MILITARY-INDUSTRIAL COMPLEX

The mobilization of the American nation after the attack on Pearl Harbor in 1941 renewed Fuller's willingness to be of service. A global war required a global analysis, and Fuller tried to find a universal language of analysis.

His point of departure for investigating nature's hidden energy system was rather mundane: table-tennis balls. He began by creating a polyhedron sculpture by gluing an equal number balls to all sides of a central ball. For arithmetical formulations and explanations of his table-tennis experiments, Fuller turned to the mathematician Homer Le Sourd, at the Naval Academy.[6] The force vectors on all sides of the figure were equal, and the distribution of force was thus in equilibrium, something Fuller believed was of fundamental importance. He thought he had found the basic building blocks of nature.

Fuller was concerned about the relevance of these Platonic exercises to the needs and understandings of the war effort. Using a cuboctahedron sculpture made with glue and table-tennis balls, Fuller glued a map of the world onto it and then cut it off again following the lines of the sculpture, thus creating a map all parts of which were equally represented. Having worked for some years as a science journalist for *Fortune* and *Life* magazines, he was able to secure the first publication of the map in *Life* as a cutout-and-glue exercise. The map appeared in the March 1943 issue, which became the bestselling issue of *Life* to that date. Thousands of American families cut out the map and arranged the pieces to form different world maps.[7]

Many innovative world maps were produced at the time, reflecting an increase in the use of the airplane; and it is in this context, as Susan Schulten has argued, that Fuller's work should be understood.[8] Maps are representations of power, and readers were able to arrange the pieces of Fuller's map so that the "North Pole layout," "Mercator's World" (around the equator), the "British Empire," "Hitler's Heartland Concept," or the "Japanese Empire" would emerge. The effect of this exercise was a sense of global understanding of the war seen from different political perspectives. The version Fuller later copyrighted and sold in 5 million copies placed the United States, Canada, and the Soviet Union together as one group at the center of the world. It focused on the air and ocean power of the Allied forces, and he consequently called it "The Dymaxion Air-Ocean World Map."

The map was a creative application of Fuller's spherical trigonometry. Together with his journalism, it provided him with enough money to spend two years on further research into the geometry of energy. The result of his calculations came in the summer of 1949, when students at the Chicago Institute of Design helped him construct his first dome. The institute, it should be recalled, was based on the teachings of its founder, László Moholy-Nagy, who taught his students to base their designs on scientific laws. One of his students was Peter Pearce, who based on his years at the institute published *Structure in Nature Is a Strategy for Design* (1978), which reformulated Moholy-Nagy's bio-technique for a new generation of designers.[9] The main principle behind the dome was thus in line with the students' syllabus. Fuller followed the principle of ecological synergy, namely, that the energetic strength of the entire construction is greater than the sum of the strengths of its parts.

The virtues of the dome technology were that it was lightweight, structurally strong, easy to build and could be adapted to meet clients' needs. Fuller soon found himself busy designing domes for clients like the U.S. Air Force, the Marine Corps, and the Department of Commerce. The Air Force's Strategic Air Command built the Distant Early Warning Line, or DEW Line, between 1954 and 1957. It was a series of radar installations along the northern part of Canada and Alaska designed to give an early warning against surprise airborne attacks from the Soviet Union. Fuller became the principal designer charged with sheltering the radar installations. The globe-shaped domes—or "radomes," as he preferred to call them—made of fiberglass and plastic, were ideal for the heavy weather and arctic temperatures of their locations. The computer surveillance systems of the DEW Line would later reemerge in Fuller's own designs for managing the earth.

Another of Fuller's major customers was the U.S. Navy, which placed a large order of domes as shelters for their marines. Assembled on an aircraft carrier, they could be lifted ashore by helicopter; thus they could quickly provide shelter for advancing forces even under atomic bombing. "It means a new era in Marine shelter," the *New York Times* reported in a photograph accompanied by photographs of Fuller in naval bliss.[10] These domes, which were to replace the traditional tents, were easy to build and cheap enough that they could be left behind if necessary. Indeed, the "Kleenex Dome," made of paperboard, was designed to be abandoned as troops moved forward. The dome was also part of "the standard of living package" that Fuller developed with students at the Institute of Design in Chicago for civilians fleeing cities to "decentralized communities" in the event of atomic warfare.[11]

Over the years these military projects became a major source of income for Fuller, whether through patents or through his firm, Geodesics, Inc. Steady royalties financed his various research projects and even allowed him to purchase a new boat, the motorship *Nagala,* so that he again could cruise as captain on his own ship. The various domes were appreciated by the military forces, which still use them.

Fuller, according to the historian Alex Soojung-Kim Pang, was a cold-war designer and a defender of American cultural values.[12] This is apparent in his numerous projects for the Department of Commerce, who initially asked him to make a dome for the U.S. pavilion at the

1956 International Trade Fair in Kabul, Afghanistan. While the Soviets and the Chinese worked on their pavilions for weeks, the U.S. pavilion was erected in just forty-eight hours. The pavilion arrived in one plane, and it took only one engineer and some untrained Afghans to complete the job. The Fuller dome thus came to represent American efficiency, military might, technological know-how, commercialism, and popular appeal, hosting a record-breaking number of visitors. The success in Kabul made domes popular at the Department of Commerce, which in the following years ordered a series of them for various fairs and expositions around the world, including one for the 1967 World Expo, in Montreal.

Fuller's fame (as well as his royalties) grew with each order, and by the mid-1960s he was one of America's most celebrated designers. No fewer than three hundred thousand domes for all kinds of purposes were built before he died. The culmination in terms of volume came with his proposal in 1960 to build a superdome over Manhattan (from Twenty-second Street to Sixty-second Street and from river to river).[13] Its purpose was to reduce the amount of energy required to heat and cool that area exactly eighty-four-fold. The military were interested in such superdomes, which could shelter troops in harsh and remote areas. This, at least, was the idea of the Russians, who, according the Fuller, contemplated constructing a dome two miles in diameter that could contain "Arctic cities" and be assembled in six months with the help of helicopters.[14] Whether or not this was true, the fact that such ideas circulated indicates that the superdomes could play both a military and a colonizing role.

ON THE ROAD: LECTURING ON DOOM AND DOMES

To mobilize more dome research and construction, and to tell the world about world problems, in the early 1960s Fuller embarked on a lecture tour that would last for more than a decade, in some ways for the rest of his life, demanding extravagant lecture fees of fifteen hundred dollars (plus money for travel and accommodations).[15] His fifty-one-page curriculum vitae, which he made into a booklet in 1975, is evidence of an unusually hectic schedule of travel to cultural institutions,[16] universities and other academic institutions, and design schools around the world. The 38 honorary doctorates, 27 awards (including the gold medal from the Royal Institute of British Architects), 26 patents, endless publications and lectures, and more than 30,000 citations should make any aspiring designer blush (which probably was his intention). The extensive lecture series resulted in no fewer than five books. All of this is surely evidence of a significant audience. Fuller's travels also gave him a personal sense of the earth as a whole, evident in his writings. The sense of managing the earth as the captain of a ship was reflected in Fuller's lecture tours, his geometry of energy, technology of his Dymaxion housing, and his veneration of the navy.

In a telling example of his status, in 1964, on the cover of *Time* magazine, Fuller's bald head was portrayed as a dome, and he was referred to within as "the dymaxion American," "the greatest living genius," "a classic American individualist," and "the first poet of technology."[17] That within a period of just five years no less than three biographies of Fuller were published is further evidence of his growing fame.[18] He became a sort of voyaging professor, with visiting appointments at 410 universities and colleges around the world (mostly in the United States). A visit might last for a couple of days, a summer, or a semester, during which time he would train students in the geometry of energy and in dome technology before leaving for the next campus.

The impending dangers that threatened the future of humankind, according to Fuller, were population growth and energy consumption, which would lead to all kinds of social and environmental ills. Fuller opened lectures to students, public appearances, and his books with this observation. The world had serious problems to be solved, and the architect-engineer had possible design solutions. *"One third of the human family is now doomed to premature death due to causes arising directly from inadequate solutions to the housing problem,"* Fuller argued in 1963, pointing to figures with dramatic curves showing the increases in human population and energy consumption.[19] The trend for the next twenty years pointed toward an imminent crisis for the human family if the current energy and housing policies were not changed. A bleak scenario of population collapse about 1972 was what his audience should anticipate if proper action was not taken immediately.

At the time of Fuller's lectures there was a general scare of atomic apocalypse, and his doomsday predictions should be read in this context. To question the future of humankind was not inapt, and Fuller's language and ideas fed on this gloomy mode of reasoning. Yet Fuller differed from the atomic doomsayers in that he provided his audience with apolitical, hands-on solutions that everyone could take part in. His practical engineering solutions were within acceptable sociopolitical boundaries.

"Nature's exquisite economy and effectiveness" served Fuller as a model for how to effectively produce shelters for the growing population.[20] The geometry of energy that Fuller used to construct domes was, in his view, the basic order of nature's ecology. Houses constructed according to the geometric order of nature would be more energy efficient, use less materials, and thus be better for the environment.

Only through radical means could the problems facing the earth be solved. Ideally, Fuller argued, scientists in collaboration with engineers and architects would be at the helm of a new "Energy-Borne Commonwealth of Humanity," steering and managing the use of energy around the world. The right movement of energy and people could thus produce a new constellation with a "Pan-American Plan on Intercontinental Cooperatives" that would secure peace in the world.[21]

Fuller presented his managerial views in perhaps his most famous lecture concerning Spaceship Earth, which he delivered in different versions starting in 1964. The National Aeronautics and Space Administration had just successfully completed its Mercury program, and it was gearing up for the launch of the Gemini spacecraft (a forerunner of the Apollo program). Space travel was in the news and on people's minds. Fuller appealed to this public interest by pointing out that all people were space travelers on a common voyage with the earth: "I've often heard people say, 'I wonder what it would feel like to be on board a spaceship,' and the answer is very simple. What *does* it *feel* like? That's all we have ever experienced. We are all astronauts."[22] The image of the earth as a huge mechanical ship traveling in space had obvious public appeal.

How to keep the ship on a steady course was the main topic of the Fuller's lectures, and the title of the lecture booklet, *Operating Manual for Spaceship Earth* suggests the engineering and managerial answer to the question: good management would require comprehensive knowledge about Spaceship Earth. According to Fuller, a state-of-the-art computer could monitor the state of the earth. Instruments in the pilothouse on a naval vessel served as the model for what he envisioned to be the chief tool for steering Spaceship Earth. This computer would be so powerful that it would ultimately replace human capacity: "Man is going to be displaced altogether as a specialist by the computer," since it would contain complete knowledge. With such a computer, comprehensive designers could act wisely. "So, planners, architects, and engineers take the initiative. Go to work," he urged, telling them to build the computer and take responsibility.[23] The planners, architects, and engineers were the only true naval officers of Spaceship Earth. They were capable of guiding the leatherneck "human family."

For Fuller, these ideas were not utopian fantasies but realizable projects he set forth to broach at Southern Illinois University. During his lengthy residency there as a research professor, he was able to work on interdisciplinary projects, such as his program for managing the earth. He believed that students needed to be taught the art of "World Planning," and he initiated a series of classes designed for this purpose.[24] It was of foremost importance that the students receive an interdisciplinary education covering all the economic, technological, and scientific aspects of world planning. In his curricula for design students Fuller emphasized the importance of a comprehensive education: "The design scientist would not be concerned exclusively with the seat of a tractor but with the whole concept of production and distribution of food," he argued forcefully.[25]

The problem with this generalist and interdisciplinary approach was the risk of losing the focus of design studies. Fuller therefore suggested "a two-hundred-foot diameter Miniature Earth" as the chief organizing tool. "This Minni-Earth could be fabricated of a light metal trussing. Its interior and exterior surfaces could be symmetrically dotted with ten million small variable intensity light bulbs and the lights, controllably connected up with an electronic computer."[26] The blinking lights on the "Geoscope," as he sometimes called it, were connected to a supercomputer running programs and projections about the earth. This knowledge was to enhance world planning. The statistical information displayed on the Mini-Earth could be used to educate the larger public about the current state of the world.

In a similar proposal for a Geoscope outside the United Nations Building in New York, Fuller argued that the human population should be plotted in red, and its dramatic growth should be displayed with the help of computer simulations as expanding or retracting color blobs. "You would see the glowing red mass spreading north-westward around the globe like a fire," he told his audience.[27] This creative use of statistics would be perceived as particularly dramatic by Americans because of the use of the chief Communist color, red, to show population growth. This use of color may illustrate how challenges to the political order translated into an ecological threat to nature itself in Fuller's work.[28]

Though the Mini-Earth was never built in New York, the project gained momentum at other locations, particularly at Southern Illinois University. In 1969 Fuller launched a grandiose, $22 million plan for a "World Resources Simulations Center" equipped with a dome (with a 5/8 sphere) measuring four hundred feet in diameter covered with electric lights driven by state-of-the-art supercomputers. It would "in effect [be] a World Brain," freeing the "mind from occupations of brain slavery," he told the president of the university.[29] Fuller's use of vocabulary from the civil rights movement as well as from H. G. Wells's book *World Brain*, published in 1938, is evidence of Fuller's rhetorical skills, as well as of the influence of Wells's eco-utopian program of the 1930s.[30] The result of Fuller's proposal was a somewhat scaled-down but still significant Simulation Center, built at the university's Edwardsville campus. The building (which now houses a religious center) was designed as a planetarium, so that when students looked out from the center of the building, they could see all the continents of the world painted on the dome. It was a microcosm of the world. The program's director would soon write enthusiastic reports about the virtue of a building that functioned like a control room for Spaceship Earth, much like "large scale displays, such as those of N.A.S.A. in Houston, and those of NORAD [North American Aerospace Defense Command]."[31] The building was thus designed as a command center for planning and steering the world toward a better future.

PLAYING THE WORLD GAME

The synergy of construction techniques and computer engineering was at the core of the Simulation Center. It was inspired by the British and American military "war games seeking the most effective means for controlling the world."³² In these games students learned to use scenarios to plan how to save the environment and to save the human population from itself. When the World Game was played at the New York Studio School of Painting and Sculpture, for example, Fuller led what in effect was a military war game to save Spaceship Earth from eventual destruction using Dymaxion Air-Ocean World Maps, which were placed on the walls just as one might find in a real war room.³³

Fuller boldly told his university superiors and his audience that he had been trained to carry out war games at the U.S. Naval Academy.³⁴ Through such games one could compare different scenarios to find the most cost-effective approach to a given situation. Making reference to the economic game theories of John von Neuman and Oscar Morgenstern and the practical experience of military planners, Fuller testified before the U.S. Senate about the virtue of computer simulations of nature's economy in the format of a grand World Game.³⁵ In a subsequent essay about the game, he emphasized the pedagogical importance of developing different scenarios for world population growth.³⁶ Some will "push buttons of Armageddon," he reasoned, but he was "betting that the earthians will *wake up and win*" in the end after playing the World Game again.³⁷

That Fuller's followers woke up after playing the World Game is evident in the reports by students and the volunteers who organized them. It was played at different scales and locations as Fuller visited different campuses and schools. A Geoscope like the one at Southern Illinois University was helpful but not necessary for playing the World Game. Those players who did not have access to a Mini-Earth would build a "game room" (analogous to a war room), with Dymaxion Air-Ocean Maps on the walls. After Fuller lectured for a day or so, his local followers (such as resident professors or graduate students) would organize the players (usually students) into different research groups. Fuller would travel on to the next campus, while the students would imagine themselves as the captain of a troubled Spaceship Earth.

The players would investigate past, current, and future needs for a particular natural resource. They would put together a material history of, say, oil or solar energy and come up with different scenarios for the future based on different technological inventions and different uses (or more often abuses) of the resources they were researching. Typically, after a few weeks the groups would get together and illustrate the results by placing markers on the maps. The World Game started with comparative analyses of the results from the different groups. This process of establishing comprehensive knowledge about Spaceship Earth was partly analytical number crunching, but it could also take the form of spiritual "travel through the minds of the others in the room."[38]

Various mystical exercises and meditations were frequently practiced to evoke the sense of being an astronaut in outer space seeing the world as a whole. No means for obtaining the captain's necessary overview of "the human family" and the earth was left untried. One player, for example, recalls the excitement of comparing scenarios for energy and labor: "I've learned lately that news has a sort of a sex life. Hegel said the same. If you take two items of antagonistic bad news and point them at each other, or *let* them at each other, you may get some surprising good news."[39] The lesson learned was that joblessness in the future could be prevented if those currently unemployed could be given the task of inventing new technologies for saving energy, thus also preventing an energy crisis. Fuller would usually return to attend the concluding sessions of the game, carefully taking notes for his future appearances and publications.

In Fuller's view, creativity, exploration, inventiveness, and play were the crucial ingredients of human survival. In his repeated critique of the Malthusian doctrine on human population dynamics (at the time advocated by Paul Ehrlich in his *Population Bomb* of 1968) as a "myth," Fuller would point to the important role of human inventive powers and claim that the World Game "demonstrated beyond question that the Malthusian doctrine is fallacious."[40] The game would inspire a whole set of environmentalists, including some Malthusian opponents of Fuller, such as the Club of Rome in its report *The Limits to Growth* in 1972, to predict the earth's behavior through computer simulations.[41] As John Markoff has shown, the people who came to shape the personal computer industry were among these early world gamers.[42]

THE CAPTAIN OF SPACESHIP EARTH

For all his cold-war futuristic and technocratic visions, Fuller was popular among counterculture enthusiasts and environmentalists of the 1970s. At first glance, Fuller's navy-inspired technocracy would appear to have had little in common with the antiwar and proenvironment agenda of hippies. But Fuller's vision of "the human family" being guarded by elite scientists and avant-garde intellectuals met with wide approval within the movement. Officers and a captain were required to guide Spaceship Earth, a role many alternative-oriented students and intellectuals found appealing.

There are numerous accounts of Fuller's support in the counterculture. For example, Fuller's insights were a chief source of inspiration for the *Whole Earth Catalog*, which since its first publication in 1969 has been one of the most popular practical handbooks for alternative lifestyles, complete with detailed explanations of how to build an ecological home as a dome. Fuller's construction techniques were used in key counterculture buildings, such as in the ultra-hippie Drop City in Colorado, where the military survival technique of doing more with less found its counterpart, thanks to Fuller, in energy-saving buildings that were "to be the easiest, most efficient, with least cost."[43] Compilations of extracts from his books and lectures, as well as videos, were made to boost interest in his thinking among the new wave of spiritualists, eco-worshipers, and other "children-of-the-earth."[44] Further evidence of the wide appreciation of his thinking is the many biographies that emerged in the 1970s. In these new biographies his navy past, the patronage from the military-industrial complex, and support from the Department of Commerce received minimal attention. Instead, he was hailed as a "Dymaxion Messiah" providing futuristic, ecological, and mystical prophesies for hippies.[45]

Fuller also met with head-on resistance from a movement marked by a rich diversity of political, spiritual, and scientific world-views. The more politically oriented leftist activists, for example, thought design revolution and ecological reasoning had a "paralyzing" effect on the revolutionary potential of the counterculture.[46] Fuller was, however, cheered by hippies concerned with spiritual thinking, environmentalism, and alternative ways to organize society. Among the environmentalists there was a fundamental tension between followers of Fuller who believed in new technologies and design and those who thought the path forward lay in a new moral and political approach toward the natural world. Acute debates between these groups often boiled down to whether one agreed with Malthus's analysis of the problem of population growth. Garrett Hardin was one of the key debaters in this discussion. He too understood the earth as a spaceship, but as a Malthusian he argued that the ship had limited "carrying capacity" and that human conduct consequently was ruled by "lifeboat" ethics.[47] A new, ecologically driven politics, ethics, and lifestyle were urgent, he argued, while those technological inventions that Fuller considered to be of primary importance were of secondary importance to Hardin. Whether one agreed with Fuller or Hardin, these Spaceship Earth debates pointed to the need for a captain.

Fuller's followers saw him as such a captain. The fact that Fuller's military rhetoric of world management found supporters among environmentalists is evidence of a movement lending its ear to authoritarian language. Fuller argued that architects and designers one day should "take over and successfully operate SPACESHIP EARTH,"[48] and his many followers were not unsympathetic to the idea. In reality, however, few put their time and efforts into organizing a governing body to operate the world according to Fuller's manuals for Spaceship Earth. The reason was quite simple: Fuller was so radical as to believe that politics soon would be a thing of the past. He and his followers were promoting an informational and not a political program for steering the spaceship and consequently put their hopes in the idea that "popular pressure" from players of the World Game would "gradually force world politics to yield to the computer-indicated, mutually-beneficial world programs."[49] The fact that the standard compilation of Fuller material made by his followers was modeled on the ideals of the seventeenth-century Enlightenment encyclopedias points to a program of pure reasoning and information rather than politics: "If humanity succeeds," according to Fuller, "its success will have been initiated by inventions and not by the debilitating, often lethal biases of politics."[50]

Fuller told architectural students that "politics will become obsolete" by the year 2000, if only designers could be in charge.[51] To readers of *Playboy* he would describe utopian new cities on the moon chemically free of politics. "Take away all the inventions from humanity and, within six months, half of us would die of starvation and disease," he argued. "Take away all the politicians and all political ideologies and leave all the inventions and more would eat and prosper than do now [in 1968]."[52] Fuller was in effect arguing for relegating politics to the role of a "secondary housekeeping [task],"[53] while assigning a leadership role to designers, whose task would be to invent technologies and construct architecture that would put the world on the right track. In Fuller's visions the political realm would fade away and ultimately be replaced by an enlightened regime of designers and technocrats.

The idea that politics, with its complex bureaucracy, would fade away carried some appeal for those inclined to Marxist reasoning, though there were deeper disagreements about the dialectics of history, politics, and the role of revolutions. Contrary to Marx, Fuller argued that history was driven by inventions, that political activity was unnecessary, and that the only important revolutions were those of design. Fuller consequently responded to the growing radicalism among students by emphasizing the revolutionary possibility of design. A "design-science revolution" was needed, he told students at Southern Illinois University. "This revolution is trying to articulate itself everywhere. It gets bogged down by political exploiters of all varieties." He argued that students involved in politics were wasting their time on a system that was incapable of offering a fair distribution of resources. He had a similar message for those trying to stop the warfare in Vietnam: "They [the peace activists] will have to shift their effort from mere political agitation to participation in the design revolution."[54] His project of ecological enlightenment aimed to take attention away from politics. The ecological movement as it was expressed in Fuller circles was thus undermining not just the civil rights movement and the Vietnam peace efforts but all political culture and activity.

A series of Fuller enthusiasts devoted considerable time to trying to make the ecological design revolution happen. In his only technical book, *Synergetics* (1975), Fuller laid the foundation for an ecological design revolution for architecture based on principles of geometry and energy. A series of technical dome studies, as well as practical "how-to" manuals for dome building written by his followers further prepared the ground for a design revolution.[55] A notable example of a design revolutionary is Victor Papanek, who in *Design for the Real World* (1971) argued that "design can and must become a way in which young people can participate in changing society."[56] Architecture and design were to him means of changing the relationship between humans and nature. While politics divided society, technologically responsible design could help to heal social and ecological relations. In short, design could revolutionize the living world.

Despite all the efforts, in 1981 an aging Fuller believed that Spaceship Earth was still "moving ever deeper into crisis" and that the optimistic scenario from 1975 of getting the ship on its right energy course by 1985 was far from being achieved.[57] His last writings are filled with doom and gloom for the earth, attributing the sad state of affairs to an invisible global economic conglomeratic monster called GRUNCH (for "Gross Universe Cash Heist").[58] The quality of these conspiracy theories, published posthumously, is perhaps best understood in the context of the ever-growing demand for information about Fuller's ideas. Since his death in 1983 a steady stream of hitherto unpublished manuscripts, reprints, edited collections, coffee-table books, art catalogs, and biographies have reached the design market. Tellingly, one book containing previously unpublished material includes an item by Fuller written in 1950 about a Noah figure designing a new ark to take the prepared away from a doomed Earth.[59] More recently, the Whitney Museum of American Art in New York produced a show honoring Fuller that made no mention of his naval technocratic solutions to the ecological crisis.[60] Today the market for items related to Fuller seems to be insatiable. On the Web there are thousands of links, including an elaborate memorial and research Web site maintained by his heirs, a retail outlet for Fuller products, a domes-for-sale store, a Bucky entertainment site, and, for readers anxious to try the ultimate power trip, opportunities to play his World Game online.

What is remarkable about most of this later material is the uncritical praise of Fuller. There is no hint of mutiny among the officers of Spaceship Earth, who loyally obey the will of their late captain. This should not be surprising, since there is hardly a tradition for self-criticism in military culture, and Fuller, as shown above, took every opportunity to fashion his environmentally concerned audience in the image of the naval community. He also tried to undermine and discredit the political realm, and since it is arguably within this realm that debating takes place, one should not expect a culture of critical discourse to emerge among his followers. Today one can observe the result: a closely knit Dymaxion community of dome builders and World Game players patiently preparing for a design revolution that will set Spaceship Earth on its right course.

Fig. 1. "Save waste paper!" Advertisement for the Container Corporation of America from 1942. Designed by Herbert Bayer, courtesy of the Artists Rights Society (ARS), New York/Bild-Kunst, Bonn.

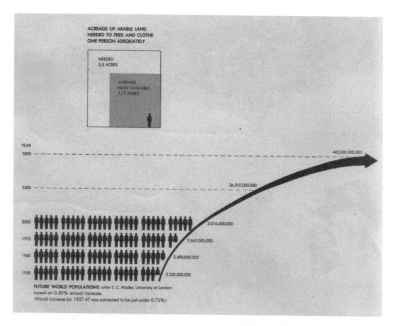

Fig. 2. "Future World Populations," from Herbert Bayer, *World Geo-Graphic Atlas: A Composite of Man's Environment* (Chicago: Container Corporation of America, 1953), 280. Courtesy of the Artists Rights Society (ARS), New York/Bild-Kunst, Bonn.

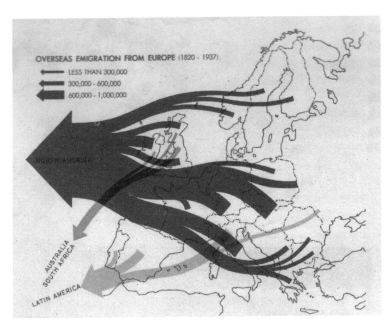

Fig. 3. "Overseas Emigration from Europe (1820–1937)," from Herbert Bayer, *World Geo-Graphic Atlas: A Composite of Man's Environment* (Chicago: Container Corporation of America, 1953), 191. Courtesy of the Artists Rights Society (ARS), New York/Bild-Kunst, Bonn.

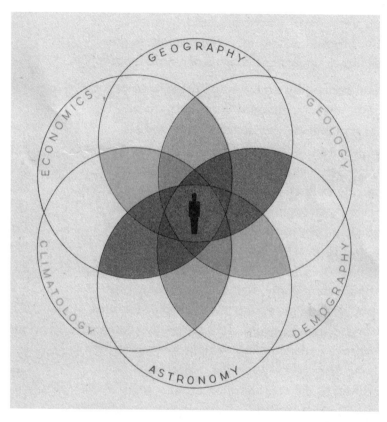

Fig. 4. "A Composite of Man's Environment," from Herbert Bayer, *World Geo-Graphic Atlas: A Composite of Man's Environment* (Chicago: Container Corporation of America, 1953), 1. Courtesy of the Artists Rights Society (ARS), New York/Bild-Kunst, Bonn.

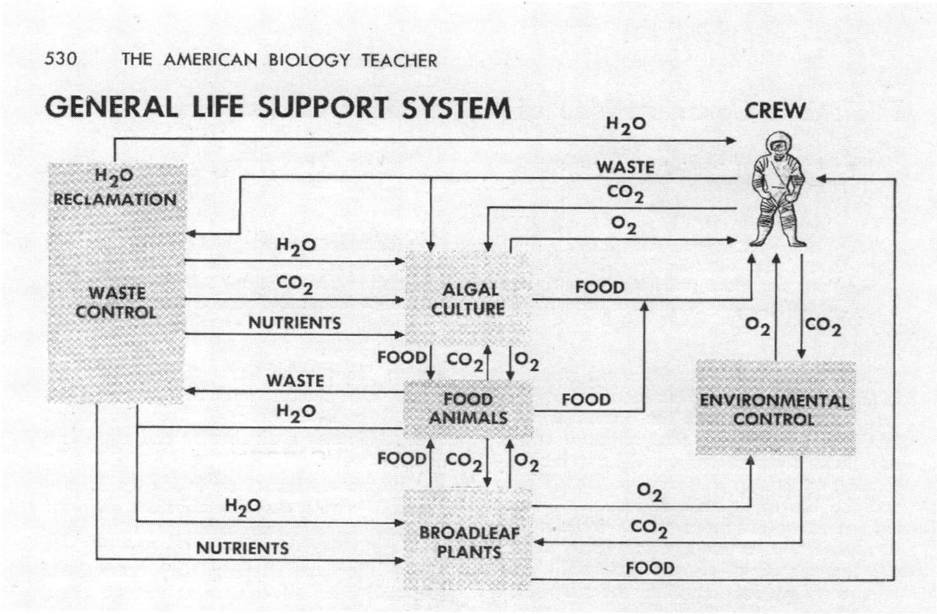

Fig. 5. "General Life Support System," in S. P. Johnson and J. C. Finn, "Ecological Considerations of a Permanent Lunar Base," *American Biology Teacher* 25 (1963): 530. Courtesy of the National Association of Biology Teachers.

Fig. 6. Illustration of a space colony in John C. Fletcher, *Space Settlements: A Design Study* (Washington, DC: National Aeronautics and Space Administration, 1977), 34. Drawing by Don Davis, courtesy of NASA.

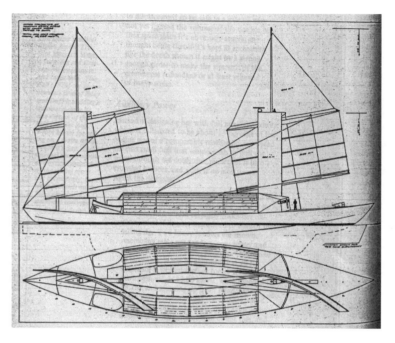

Fig. 7. Illustration of a New Alchemist ocean ark in John Todd, "Ocean Arks," *Co-Evolution Quarterly* 23 (1979): 52. Courtesy of John Todd.

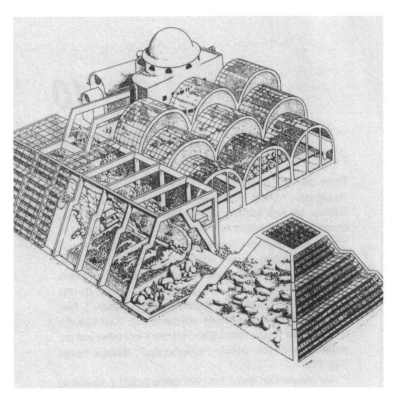

Fig. 8. Drawing by Sheila Manion-Artz in Dorion Sagan and Lynn Margulis, *Biospheres: From Earth to Space* (Hillside, NJ: Enslow, 1989), 78. Courtesy of Lynn Margulis.

6. THE ECOLOGICAL COLONIZATION OF SPACE

"Imagine that Mars is a Utopia in which there is complete trust, total harmony, no selfishness and no deceit. Now imagine a scientist from Mars trying to make sense of human life and technology [on Earth]," Richard Dawkins urged readers in *The Extended Phenotype* (1982). He attributed this Martian outlook to the "pop-ecology literature" of his antagonist James Lovelock and to the followers of Lovelock's Gaia thesis.[1] The image of an ideal community on Mars was not accidental, as leading ecologists in the 1970s were investigating how to construct colonies on Mars and the relevance of this research for understanding ecosystems on Earth. Edward O. Wilson, for one, advised readers of his *Sociobiology* (1975) to view life on Earth as "a perceptive Martian zoologist" would.[2]

In this chapter I investigate what ecologists sought to do on Mars and what the Martian perspective meant for their understanding of life on Earth. It is a history that originates in military research into constructing self-sufficient, closed ecological systems within submarines and underground shelters. With the space program of the 1960s, leading ecologists suggested using this know-how to construct closed ecological systems within space capsules, ships, and colonies. Their research into the number of astronauts that could ecologically be accommodated in a spaceship was subsequently used to determine the carrying capacity of Spaceship Earth. In the 1970s, environmental ethics became a matter of trying to live like astronauts by adapting space technologies such as bio-toilets, solar cells, recycling, and energy-saving devices. The technology, terminology, and methodology developed for the ecological colonization of space became tools for solving environmental problems on board Spaceship Earth.

HUMAN ECOLOGY IN SPACEFLIGHT

Cabin ecology was the term used in astronautics in the late 1950s to describe the environment inside a space vehicle. It was believed that the environment of a space cabins should be as close as possible "to the environment found on the surface of the earth."[3] A body of about one hundred engineers and scientists worked on the development of systems such as those for circulating air, water, and food. This research on cabin ecology was sponsored by the military, which considered it to be of key importance for the construction of submarines, atomic shelters, and environmental planning.

In 1958 submarine development was the focus of about two hundred scientists sponsored by the American Institute of Biological Sciences who were working on various cabin ecological designs. One of them, the biologist Jack Myers, at the University of Texas, had for years studied "the use of plants to regenerate air in a closed ecological system, such as that of a space cabin." His research was used, according to a report published in the journal *Missiles and Rockets*, by submarine engineers to improve the "space flight under sea." The use of "ecological systems for underwater vehicles . . . [in] the design of space cabin ecological systems" later became the norm.[4] It was this naval interest in ecological engineering that triggered Buckminster Fuller's thinking about Earth as a capsule.

The construction of shelters was also tied to cabin ecological designs. When President John F. Kennedy outlined in his program for a nationwide civil defense effort in the autumn of 1961, it threw the country into a shelter-building mania. All over America people began to build shelters according to design handbooks made available by the Office of Civil Defense. One of OCD's technical manuals, *Shelter Design and Analysis* (1965), typically states that "problems of habitability in closed ecological systems have been solved in the past with excellent results. Notable examples of such systems are submarines and space capsules."⁵ The manual simply applied closed ecological systems developed for space cabins to large and small underground shelter designs. Inside these underground cabin ecologies people were to survive for years, months, or weeks (depending on their military importance). Exploring the role of these cold-war shelters, Tom Vanderbilt noted their importance in the history of closed-ecosystem research.⁶

Military environmental planners of human settlements also took interest in cabin ecology. "A good decentralization program would . . . reduce the target attractiveness" in atomic warfare, a military analyst typically thought.⁷ Effective atomic defense required dispersion of the population to the countryside, where people ideally would live on self-sufficient small farms while also learning to survive in wilderness. The ultimate program of decentralization would be to place humans in space to secure the survival of the human species in case Earth was destroyed in an atomic apocalypse. Although the word *ecology* was widely used in this military research, few professional ecologists were actually engaged. This would change in 1962, when the Ecological Society of America arranged a session to discuss the emerging science of space ecology at the annual meeting of the American Institute of Biological Sciences.

In May 1961 President Kennedy announced his belief that the nation should commit itself to sending a spaceship to the moon and returning it safely back to Earth. In view of this ambitious program, and the research money that followed it, the time was ripe for ecologists to get involved. When it invited leading space engineers and ecologists to share a session in 1962, the Ecological Society was trying to establish a closer liaison between space researchers, military engineers, and natural ecologists. The 1962 session led to three conferences at Princeton University, in 1963, 1964, and 1965, on the topic "Human Ecology in Space Flight," arranged in collaboration with the Office of Naval Research and the National Aeronautics and Space Administration, or NASA.⁸

The brothers Eugene P. Odum and Howard T. Odum were two leading ecologists who jumped at the opportunity to put ecology in the service of space exploration. Historians have discussed how in the 1950s and 1960s the Odum brothers, thanks to patronage from the Atomic Energy Commission, came to the forefront of the field by introducing energetic systems theory to ecology.⁹ By diagramming the flow of energy in the natural world as input and output circuits in a cybernetic ecosystem, they provided ecologists with new methodology and research techniques. Their social program was to bring human activities into balance with the ecosystem through natural, social, and technological engineering. At the University of Georgia this was a personal issue to Eugene Odum, as the Ecology Building, where he worked, was designed to reflect the faculty's ecosystem research.[10]

The Odum brothers were firm supporters of the ecological colonization of space throughout their lives. Eugene Odum thought building space cabins was simply a question of taking "a little piece of this biosphere . . . and try[ing] to build a wall around it so that it [would] be materially closed but not closed to energy flux."[11] Howard Odum agreed. It was possible to support humans in space by constructing "a *climax (steady state) ecosystem of many species*" within the space cabin with a "carrying capacity" of a few astronauts.[12] The term *carrying capacity* was first used by Mark Twain in 1883 to describe the maximum load of people and goods that a steamboat could carry, and since then it had been used mainly in shipping.[13] The Odum brothers used the term correspondingly to refer to the number of astronauts a spaceship could support and still maintain itself. Close management of the population dynamics of species on board would be of paramount importance for the ship's survival. The astronauts would have to live in harmony with the spaceship, and this would be equally important in the case of humans on board Spaceship Earth.

Engineering a viable ecosystem in space would be the chief professional challenge for ecologists. According to Eugene Odum, the solution was to "combine natural components with mechanical shortcuts" in meeting the cabin's need for solar energy, recirculation of sewage, air, and water, and production of the crew's food (such as algae or slugs).[14] Other ecologists suggested that microorganisms grown in human urine could be a source of nutrition. They regarded the astronaut as an integral part of the closed ecosystem. In the words of Jack Myers, "The human goes into space, not as a passenger, but as an essential part of the instrumentation needed for a particular mission."[15] This instrumentalist view of human agency as serving the higher mission of the spaceship was later projected into ecologists' writings about the ethics of carrying capacity for humans on board Spaceship Earth.

The space cabin was to be a self-sufficient and stable ecosystem. "Complexity lends stability," argued the marine biologist Bernard C. Patten, who saw a "connection between information theory and ecological theory" with respect to the "self-organizing capabilities" of both systems.[16] The idea of using plankton and algae as food resources was justified by the possibility of synchronizing the computer design of the spaceship with the statistical behavior of plankton communities. The robustness of both systems would depend on the complexity of information circuits and the number of species. Different species within an ecosystem would feed on one another and thus maintain the stability of the system. The space cabin should therefore be equipped with high-tech, computer-driven air conditioning, composting, bio-toilets, algae-assisted sewage rinsing, and solar cells. These biotechnologies were to secure human evolutionary expansion and adaptability in space and would later be regarded as soft-tech, eco-friendly technologies necessary for human survival on board Spaceship Earth.

Ecologists also created designs for a lunar base. To be situated forty feet below the lunar surface, such a base was imagined as a closed ecosystem of 250,000 feet with a carrying capacity of twenty-five astronauts. The proposal was supported by a "general life support system" complete with charts for the circulation of oxygen, carbon dioxide, nutrients, waste, plants, animals, and astronauts (Fig. 5). Eugene Odum's work on climax ecology served as the methodological foundation for constructing a steady-state biotic community on the moon. A larger ecosystem was better than a small one, he argued, since a "diverse system is safer" owing to its robust biotic complexity.[17] The lunar base was to circulate all its material resources, an aim that later became the architectural ideal for ecological buildings on board Spaceship Earth.

The lunar base was imaginative, though NASA was looking for simpler solutions for the first flights. General agreement emerged from the Princeton conferences that it was unnecessary to build the self-maintained ecosystem ecologists imagined for a flight of only a couple of weeks. It would be simpler to rely on food storage and chemical rinsing of air and water than to incorporate biological systems into the spacecraft. For this reason ecologists would not take part in the engineering of the first space cabins, the space shuttle, or the space station that eventually orbited Earth.[18] At the same time it was also generally agreed that functioning ecosystems would be necessary for voyages into deep space and for lasting bases on the moon or Mars, since it was too expensive and complicated to supply such projects from Earth.

Though ecologists failed to offer NASA a workable short-term proposal for a cabin ecological system, they did succeed in providing a viable outlook for things to come in the future. Their ideas would consequently capture the imagination of science-fiction and futurist writers. The founder of the British Interplanetary Society, Arthur C. Clarke, for example, used ecological arguments as the scientific basis for the screenplay for Stanley Kubrick's film *2001: A Space Odyssey* (1968). "The [Moon] Base" imagined in the movie, Clarke explained, "was a closed system, like a tiny working model of Earth itself, recycling all the chemicals of life." He based the screenplay on human ecology in space research as well as on the biological mysticism of Pierre Teilhard de Chardin (1881–1955). According to de Chardin, a transcendent powerful "Omega point" turned evolutionary history. In the movie this Omega point was visualized in the discovery of a powerful stone signaling the coming of a dramatic evolutionary turn. Space ecology made the discovery of the stone possible. In the book version of the movie the space explorer leaves an Earth devastated by environmental ills, pollution, and alarming human population growth wondering whether Earth "would still be there when the time came to return."[19] After a stopover at a local space station, he arrives at the moon base to investigate the newly disinterred mystic stone. The moon base was built as a hothouse below the surface of the moon with a balanced ecological system securing the survival of the human race in case the earth collapsed in an eco-disaster. The last, psychedelic part of the movie plays on the idea that humans could evolve to a new evolutionary level and reach the Omega point by taking LSD. Such fantasies were intriguing, especially to the larger public who crowded the movie theaters.

Understanding the environment as a closed ecological system was also, as Kim McQuaid has shown, an effective way for NASA to sell the space age to the larger public.[20] In addition, closed systems were easy to communicate to students. If the complexity of Earth's nature could be replicated in a small space cabin, one could also replicate it in textbooks. The fact that some of the papers on space ecology were published in *American Biology Teacher* reflected the idea that cabin ecology was an ideal way of bringing the complexity of nature into classrooms. Dennis Cooke, a graduate student of Eugene Odum's, would emphasize this in his chapter to the third edition of Odum's classic textbook *Fundamentals of Ecology* (1971). He claimed that space exploration was "one of the most exciting new areas in science," generating the necessary "lebensraum" for human evolutionary expansion. By evoking this justification of colonialism, he sought to inspire new students of ecology to devote their research to space exploration. Odum agreed: he believed that the ecological colonization of space needed help from the scientific community, "and a little financial support from NASA." As Ramón Margalef noted, space ecology was "likely to draw more attention (and surely more money!) than biology."[21]

THE NEXT ECOLOGICAL FRONTIER

The ecological colonization of outer space was a technically and economically viable idea, at least if one believed the former astronaut candidate and physics professor at Princeton University Gerard K. O'Neill, whose bold visions for space colonization caught the imagination of ecologically minded thinkers in the 1970s. In 1969 O'Neill's students confronted him with his profession's entanglement with the military-industrial complex, atomic weaponry, and environmental destruction. In response he designed a course, Physics 103, aimed at studying physics that could produce peaceful solutions to the world's problems. Soon the students were engaged in calculating what it would take to build a colony in space. It was supposed to be free of military purpose, in ecological harmony, and without atomic pollution or other suspicious industrial activities and to contribute to the well-being of Earth (including the needs of the inner cities). This research resulted in two articles that appeared, after some delay, in *Nature* and *Physics Today* in 1974.

"Careful engineering and cost analysis shows we can build pleasant, self-sufficient dwelling places in space within the next two decades, solving many of Earth's problems," O'Neill argued. The idea was to use material resources on the moon to fabricate a grand space station located at one of the points of gravitational equilibrium between the moon and Earth. The station, complete with mountains, lakes, and small-town communities, was to be based on John D. Bernal's model for creating artificial gravity in space by means of rotating spheres. Moving heavy manufacturing to the moon could relieve Earth of polluting industries, and a grand space station could ease Earth's population pressure. Such a space station, he argued, was "likely to encourage self-sufficiency, small-scale governmental units, cultural diversity and a high degree of independence." It was to be located at one of five Lagrangian liberation points between Earth and the moon, a point where gravity is in equilibrium. Consequently it was named L-5 at the suggestion of George Hazelrigg, also of Princeton. It was to be an Arcadian ecological community rooted on managerial principles. The fact that the space station can be understood within both an Arcadian and a managerial tradition in ecology illustrates the difficulty of supporting this historical dichotomy.[22]

The articles raised eyebrows among physicists. According to O'Neill, however, it was not a far-fetched idea in view of the Spacelab program NASA had successfully carried out in three missions in 1973 and 1974. Thanks to a series of public appearances, he soon became a physics celebrity and would receive "thousands of letters" from the public about the space colony. Judging from the depiction of him in *Penthouse* magazine, dressed as an astronaut with the space colony in the background, his readership extended far beyond the physics community at Princeton.[23]

Former students and environmentally concerned hippies became O'Neill's most loyal supporters. They saw in him a brave professor who was willing to break with suspect atomic research to pursue eco-friendly physics. One of them was Stewart Brand, the editor of the *Whole Earth Catalog,* who used the royalties from this highly successful guide for unorthodox living to generate a research fund, the Point Foundation. Brand, the foundation, and the network of activists that came to surround him, as the historian Fred Turner has shown, moved the counterculture toward high-tech and cyberculture solutions to environmental and social ills.[24]

Brand became O'Neill's patron, covering the expenses for the First Conference on Space Colonization at Princeton University in 1975, which became a fairly technical conference designed to show the seriousness of space colonization. One paper published in *Science* argued that O'Neill's suggestion for mining on the moon was feasible and should be pursued. The system analysts Keith and Carolyn Henson presented a paper entitled "Closed Ecosystems of High Agricultural Yield," in which they argued that the space ecosystem could provide, among other things, meat from alfalfa-fed rabbits and dairy products from goats. Energized by the conference, the Hensons started a support group that for a decade promoted O'Neill's ecological space station. Inspired by the systems engineer Wayne A. Wymore, at the University of Arizona, they calculated the human carrying capacity of future space farms.[25]

John C. Fletcher, at NASA, was equally excited and organized a study group at Stanford University "to design a system for the colonization of space" in response to the "finite resources and ominous pollution" on "spacecraft Earth." The result was one of the more imaginative reports from NASA, with colored illustrations of the space colony (Fig. 6) supported by scientific figures such as an ecological diagram of the circulation of water within the capsule. The group concluded that "Space colonization" was desirable because it offered hope to humanity living in a limited world where "the delicate ecological balance of the planet" was in trouble. Space offered literally "a way out, with new possibilities of growth and new resources."[26] It would be, according to a follower, "a pollution-free world."[27]

While the Stanford group was still working, O'Neill testified before the Subcommittee on Space Science and Applications of the U.S. Congress that an investment of $178 billion for a space community of ten thousand people would be paid back in twenty-four years through the sale of environmentally clean energy beamed back to Earth. Given the limits to growth on Earth (as described by the Club of Rome's alarming report of 1972), O'Neill explained, it was of paramount importance to expand into space. "Not a penny for this nutty fantasy!" declared Senator William Proxmire, a Democrat opposed to wasteful government spending.[28] He did not sway the subcommittee, however, which recommended a 25 percent increase in NASA's budget to prepare for space colonization.

Intrigued by O'Neill's vision, NASA provided him with money to do research for his widely read book *The High Frontier* (1976). A space station could solve most of Earth's environmental ills, he argued. It could be built as a "steady state" economy in harmony with the station's ecologically engineered system, and clean energy could be sent back to

Earth from one of Peter E. Glaser's solar-powered satellites in space. Not only could space stations benefit the planet "by relieving Earth of industry and of its burden of population" but its ecosystem, with "species of animals, birds and fish in danger on Earth," would have a better chance of survival in space. The space station was thus to be understood as a Noah's Ark, taking an intact ecosystem into space, away from Earth's polluting industrialism. "Noah's passenger list" of species traveling to the space station included only those that were of "pleasure to us" and necessary for "a complete ecological chain," while "annoying scavengers," such as wasps and hornets, were to stay on Earth.[29]

O'Neill's ideas were based on Daniel S. Simberloff and Edward O. Wilson's ecological methodology for understanding the colonization of empty islands, in addition to Eugene Odum's work. The ecological "islands in space" were not, O'Neill insisted, science fiction but "depended on present-day technology, on machines which we are sure we can build within the limits of our present knowledge."[30] He would emphasize the scientific validity of his claims again and again, backing them up with calculations, tables, footnotes, and his authority as a professor in the prestigious Department of Physics at Princeton University (once the home of Albert Einstein).

Among those who were clearly impressed was the popular-science writer Tom A. Heppenheimer. In 1977 Heppenheimer published *Colonies in Space,* which was widely distributed through the Book-of-the-Month Club, the Natural History Book Club, the Quality Paperback Club, the Playboy Book Club, and the Macmillan Natural Science and Explorers Book Club. For popular-science readers it thus became the standard overview on the subject. The same audience had previously been exposed to the worrying news in *The Limits to Growth* (1972), and Heppenheimer's book was sold as a remedy to the Malthusian problem of human growth on Earth as this was understood by the Club of Rome. The technological solutions that came with space colonization, Heppenheimer argued, would solve Earth's energy problem as well as find new space for an overpopulated Earth. "The space colony" would "be a closed-cycle ecology *par excellence,*" since all living things would have to live in harmony within the artificially built ecosystem in order for the colony to survive. Solar energy would support the farming of vegetables, fruits, fish, birds, and various animals. The Edenic vision of the space colony as "a land of milk and honey" in balance as "a completely closed ecosystem" served as a contrast to the polluted, ill-managed, and ecologically unbalanced environment on Earth.[31]

The support for the high frontier of ecological space stations would grow rapidly. The governor of California, Jerry Brown, for example, announced that "ecology and technology find a unity in Space" in his speech on Space Day (named in analogy to Earth Day) in 1977. He was a firm supporter of cabin ecological research, which from 1978 was carried out at NASA under the heading "Controlled Ecological Life Support Systems."[32] Space colonies were not conceived as science fiction but as workable proposals worthy of scientific research. From the point of view of human geography, space stations were an intriguing new environment to analyze.[33] The entertainment industry also found space ecology intriguing. Disney's Experimental Prototype Community of Tomorrow (EPCOT), in Orlando, Florida, which opened in 1981, included a grand Buckminster Fuller–inspired dome named "Spaceship Earth," in which visitors could marvel at panoramas showing people living in a high-tech, ecological harmony inspired by O'Neill.[34]

In the early 1980s both O'Neill and Heppenheimer restated their arguments by drawing up visions for eco-friendly space stations, as well as moon and Mars bases. A steady stream of followers have pursued their ideas for ecological colonization of space, the moon, Mars, and beyond.[35] One of them was the ecologist and historian of science Frank B. Golley, who believed that the ethic of colonization was an inherent quality of Western culture. The environmental ethics of "the space colony," he predicted, would set the standard for ecological living "over the entire Earth."[36]

THE ECOLOGICAL CYBORG AND THE ETHIC OF THE ASTRONAUT

The ecological colonization of space involved a new ethic of conserving energy, recycling, composting, and family planning. This ecological ethic had its origin in the lifestyle of the astronaut, which implied a new set of technological, ethical, and social tools to guide humans to adopt the astronaut's way of life. Thus, the cyborg became the model for living in ecological harmony.

The future was to be organized according to human biological needs, at least if one believed the professor in biomedicine René Dubos, who in 1972 argued that "spaceship ecology" implied "entirely new technologies" for human living.[37] Numerous designs for such technologies can be found in Brand's *Whole Earth Catalog*, published in various incarnations since 1968. The catalog has been the object of an excellent study by the historian Andrew G. Kirk, who argues that it became a key vehicle for the counterculture, representing a bold attempt to "reconcile nature and the machine."[38]

The research of John McHale may serve to illustrate the type of technological solutions the *Whole Earth Catalog* promoted. He was one of the chief proponents of building a new future in outer space, a keen follower of Buckminster Fuller, and the director of the Center for Integrative Studies at the State University of New York at Buffalo. Fuller believed that the presupposition of "a non-flying-man ecology" was mistaken and that humans were destined to have a healthy ecological future in space.[39] McHale agreed. A dramatic ecological design revolution was about to change human evolutionary history, he believed.

Humans had previously "spread out horizontally into every corner of the planet," McHale argued in *The Future of the Future* (1969), but were now entering a radically new phase of spreading "into space and down to the bottom of the oceans," which signified "another evolutionary form" and the coming of the age of ecology. McHale considered the fusion of ecology and cybernetics in space and submarine technology the latest and most important shift in human history. "The organic fusion of the human organism with active cybernetic components" would create a bionic "cyborg" in ecological harmony with the new space or submarine environments, McHale thought. Images of astronauts in spacesuits, people engaged with robots, and the use of various bionic devices were to McHale examples of future ecological living. His vision was built upon the "closed ecology" of the space cabin, or what he also called "the microearth capsule."[40] His point was not that in the future everyone would live their lives in spacesuits (though some would) but that if life on board a spaceship was to be in ecological harmony, humans would have to learn to live as ecological cyborgs. Astronauts would have to adjust their lives to a host of computer-driven, cybernetic monitoring and control systems in order to steer their spaceship in ecological harmony. Humans of the future would have to make technologies for renewable energy, solar cells, recycling of air and water, waste processing, sewage management, material reuse, and other health-related technologies developed for space stations part of their daily lives. The task of ecologists would be to monitor and control all these devices, with the help of a giant computer, for the benefit of the spaceship's overall cybernetic system. In Fuller's view, the control room for the scientific, ecological steering of life in outer space would eventually look like the Mission Control Center at the North American Aerospace Defense Command (NORAD). *The Future of the Future* was well received by reviewers who understood it as an interpretation of Buckminster Fuller's thinking and as an important book in its own right.[41]

The systems designer Mike Waters was one of those inspired by McHale. He thought the human relationship with nature would gradually evolve into a "green cyborg" condition thanks to space-cabin technologies. It was an ecological elaboration of other cyborg studies imagining "incorporating artificial organs, drugs and/or hypothermia [in astronauts] as integral parts of the life support systems" of spaceships.[42]

These ecologically construed cyborgs later became a key source of inspiration for Donna Haraway's thinking about the need for humans to reconnect with the natural world. The historian of science Michel Serres was also inspired. To him the life of astronauts promised a renewal of "the natural contract" imagined by the French romantic philosopher Jean-Jacques Rousseau. "All humanity is flying like space-walking astronauts," he argued in his plea for making peace with "the fastest shuttle. The most gigantic rocket. The greatest space ship [Earth]."[43] The thinking of both Haraway and Serres, it is worth noting, would evolve in directions far from the technocracy of space age enthusiasts, as would the thinking of landscape designers inspired by cabin ecological research.

7 TAKING GROUND CONTROL OF SPACESHIP EARTH

Space exploration has created "for architects, landscape designers and city planners the conceptual basis for a wholly new approach to the design of human settlements," James M. Fish, the renowned professor of architecture and environmental design at Columbia University, noted in 1971.[1] The ecological approach to construction of space-cabin environments for astronauts linked space exploration to design. These space cabins, Fish argued, should serve as models for environmentally responsible landscape and architectural designs on Earth, an idea that was widespread in the ecologically minded design community at the time.

LIVING ON BOARD SPACESHIP EARTH

The image of Earth as a giant space cabin sailing through space with human astronauts on board came to dominate ecological debates in the late 1960s and 1970s. The framing of nature in terms of life in spaceships enabled an ecological ethic for humans on Earth modeled on the scientifically manageable astronaut.

Buckminster Fuller was probably the first person to write about the role of human ecology in space research in publications directed at nonspecialists. As the designer of a series of domes and maps used for military purposes and a keen admirer of the navy, he knew firsthand what was going on in military research. As early as 1963 he noted that "billions of research dollars" had "been applied to a closed chemical circuit of ecologic . . . living of moon-rounding men." About the same time, Fuller started using cabin ecology in his lectures as a model for understanding life on Earth, as in his *Operating Manual for Spaceship Earth* (1969), discussed in chapter 5.[2] Among the many designers inspired by Fuller were Ulrich Franzen and Paul Rudolph, former students of Walter Gropius who made drawings of futuristic cities on "spaceship earth" modeled on imagined space colonies.[3]

Fuller's lectures also inspired the economist Kenneth E. Boulding to write his influential 1966 article "The Economics of the Coming Spaceship Earth." It was the first attempt to apply cabin ecology to macroeconomics. Boulding distinguished between an unruly "Cowboy economy," with an "open" and exploitative ethic, from virtuous, "closed" economic systems, with an ethic of responsible management of the earth as a grand spaceship. The article soon became a standard reference for eco-friendly economic theories.[4]

Also inspired was Barbara Ward, an international economist at Columbia University who in her 1966 book *Spaceship Earth* used movement of energy within the space cabin to understand life on board Spaceship Earth. Like Boulding, she was no space enthusiast and believed money would be better spent solving environmental problems on Earth. She adopted the managerial ethics developed for space exploration to steer social energy. "Most of the energies of our society tend towards unity" of people, Ward argued. She thought that the United Nations was a promising center for energy steering and research into the carrying capacity of Spaceship Earth. Science-based politics was the way forward, and she organized conferences where scientists and politicians met to point the ship toward an orderly future. This reasoning inspired Adlai Stevenson, the U.S. ambassador to the United Nations and the 1952 Democratic presidential candidate, to note that "we travel together [as] passengers on a little spaceship."[5]

Spaceship Earth soon became a key term in the U.N. vocabulary, especially after Secretary General U Thant used it in connection with Earth Day in 1970. "Spaceship earth is left without central guidance and stewardship," he complained in a speech about the world's lack of commitment to U.N. leadership.[6] The undersecretary general of economic and social affairs, Philippe de Seynes, also argued that Spaceship Earth signaled a new commitment to "globalism," which, unlike "internationalism," sought to analyze the world in terms of "the degradation of the environment, the destruction of ecological balances, the limited capacity of the biosphere, the possible depletion of natural resources, the population explosion, the finiteness of the planet, and perhaps even the finiteness of knowledge."[7]

Concerned environmentalists could not agree more. In the 1970s the term *Spaceship Earth* was often used in addressing ecological issues and the urgent need for global leadership. Future "helmsmen on the spaceship Earth," one environmentalist argued typically, should base the political realm on a secure scientific footing.[8]

In 1970 Fuller's assistant McHale published a book in which he explained how to follow up Fuller's *Operating Manual for Spaceship Earth* by living within the life-support system of Spaceship Earth. *The Ecological Context,* as the book was called, was an attempt to monitor the use (or more often abuse) of the earth's energy and material resources according to Fuller's ideas. It was originally drafted as a report for the World Resources Inventory, at Southern Illinois University, where Fuller was in charge and McHale for a period served as executive director. It was crucial, according to McHale, to view humans in the "ecological context" of "planetary housekeeping" and not from "traditional political and economic viewpoints which have guided and measured his large-scale actions before." Using the space cabin as an explicit model, he described the world as a closed ecosystem with cycles of population, food, energy, and various other materials. McHale defined the human being as "an energy-converting organism" malfunctioning within the closed ecological system of the earth and thus causing a series of environmental ills. This led him to conclude that in comparison with the ideal ecosystem within the space cabin Spaceship Earth was out of balance. What was needed was an "Ecological Redesign" of the household of nature through new "housekeeping rules" overseen by specialists at the United Nations.[9] In short, the earth should be modeled on space-cabin ecology, and humanity would have to behave like astronauts in order to live in harmony within its system. As a major organizer of future studies, McHale would promote such ideas for years to come.[10]

The cabin ecologists also used the spacecraft as a model for the earth. At the American Astronautical Society conference in 1968, "Bioengineering and Cabin Ecology," for example, the opening lecture was all about the future well-being of Spaceship Earth. The "close analogy between the ecology of a space cabin and the ecology of the planet Earth" served as a point of departure for reflecting on the world's environmental and social conditions.[11] Thanks to a series of technologies for managing waste, air, food, and energy, the space cabin came to represent the rational and scientific way of living ecologically. Humans on board Spaceship Earth, by comparison, were polluting their cabin with carbon dioxide, were hardly recycling their waste, and did not generate enough energy from the sun. It was consequently urgent to transfer technology from the space capsule to Earth. This, at least, was the view of a representative from the Lockheed Missiles and Space Company, which believed technologies developed for the lunar base would be ideal for solving many of the ecological imbalances on board Spaceship Earth.[12]

Viewing Earth as a giant space cabin required a panoramic perspective, which was provided by images of Earth the Apollo spaceship sent from the moon.[13] The photos soon came to represent, as Sheila Jasanoff has argued, "emerging global norms" for taking care of a common environment.[14] The view inspired many ecologists who also used the imagined communities of future space colonies to analyze the earth. In *Environment, Power and Society* (1971) Howard Odum made the case for understanding the earthly environment and human activity in terms of astronauts' life in outer space. "The biosphere is really an overgrown space capsule, and the questions about carrying capacity [for man] are similar," he argued. He did not use the space capsule as a vague analogy or metaphor, but as an ontological claim about the world. His methodological reductionism of all biological life (including human behavior) to charts of energy circuits became the justification for his proposals for scientific management of human society. To live in harmony with the earth's ecosystem was to him a question of adopting space technologies, analytical tools, and ways of living. With his wife, Elisabeth Odum, he would point in *Energy Basis for Man and Nature* (1976) to the importance of a "steady state economy" and an understanding of "our *life-support system*" on Earth as analogous to astronauts' life-support systems in a steady-state spaceship. Likewise, as late as 1992 Eugene Odum would structure his ecology textbook about "life support systems" of Spaceship Earth in accordance with the life-support systems of the Apollo spaceflights.[15]

A similar line of reasoning was promoted by James Lovelock, who in the mid-1960s suggested a method for detecting life on Mars based on what life on Earth looked like from outer space. He developed and patented a series of detection devices that NASA bought for planetary exploration. From 1961 to 1967 he explored these as a consultant for the Jet Propulsion Laboratory, at the California Institute of Technology, receiving about six thousand dollars a year. After 1967 this engagement evolved, with contracts as an independent scholar worth on average twenty thousand dollars a year, including overhead, travel, and various costs associated with fabricating the devices. His work was quite successful, as two pieces of his hardware eventually traveled with the Viking landers to Mars in 1975–76.[16] Thus Lovelock was not only financially dependent on space exploration but had also taken part in the technical development of space technologies and human ecology in space research. The space cabin was designed as a self-regulating cybernetic system with the capability of maintaining the chemical components of the atmosphere through negative and positive ecological feedback loops that provided comfortable living conditions for the astronauts. The Gaia hypothesis he proposed in 1974 with Lynn Margulis basically postulated Earth as a giant space cabin, complete with a self-regulating system that maintained climate and chemical compositions comfortable for living organisms.[17]

There were also those who came to understand the earth in terms of a spaceship even though they did not support space colonization. The population biologists Paul Ehrlich and Richard L. Harriman, for example, fashioned every aspect of life on board Spaceship Earth in accordance with life within a spaceship. The biologist Garrett Hardin also explored the ethical relevance of fashioning the earth as a spaceship. In the early 1970s he developed a special "lifeboat ethics," framing environmental ethics for the ship in naval martial codes. His point was that the population problem on Earth was like that of an overcrowded lifeboat. A suspension of humanist ethics was necessary to keep the boat afloat, he argued, just as naval military codes might require sacrificing marines to save a ship in trouble. Lovelock agreed. Humans were to him "pollution" spreading "like a disease" threatening to kill Gaia.[18]

These anti-humanist ideas were taken seriously by environmental activists such as David Foreman, Ynestra King, and Christopher

Manes, who by the late 1980s were taking part in acute arguments on whether AIDS was Gaia's solution to population growth. Whether to give food to the starving or simply "let nature seek its own balance" became a moral dilemma. Most radical perhaps was Foreman, whose organization Earth First! championed physical destruction of things that could unbalance Spaceship Earth. Although professing pluralism and nonviolence, the organization was run like a guerrilla group ready to destroy in order to save the environment from destruction.[19]

CAPTURING THE DREAMS OF THE COUNTERCULTURE

This turn to space ecology as a beacon of hope should be understood in the context of gloomy ecological predictions for the earth. In the late 1960s and early 1970s a series of alarming reports dominated environmental debates. Paul Ehrlich's *The Population Bomb* (1968) and the Club of Rome's *Limits to Growth* (1972) caused intense debates about the future of industrial societies and ways of avoiding a global ecological collapse. Architects were also among the concerned, as demonstrated by "Designing for Survival," a special issue of *Architectural Design* published in 1972.[20] It was adorned with an image of a human skull emerging from industrial pipes, which captured the mood of the articles: humankind would face a certain death unless something was done about pollution. The task of designers, the articles in the issue argued, was to secure the biological survival of the human species by incorporating ecological principles into future buildings.

The ecological state of the world was only one of many disturbing issues, along with the cold war, the Vietnam War, violent civil rights demonstrations, and the struggle for women's liberation. In this period of questioning authority, the space program came to represent a beacon of hope for the counterculture. People of the so-called '68 generation, the historian Mark Kurlansky has argued, viewed space exploration with "tremendous excitement."[21] As discussed in chapter 6, a series of scholars, economists, politicians, and environmentalists of the period pointed toward space ecological research as a remedy for the eco-crisis. The notion of an ecological "carrying capacity" for astronauts within a spaceship was systematically used to develop carrying capacities on Spaceship Earth. Population biologists such as Paul Ehrlich, Richard L.

Harriman, and Dennis C. Pirages, for example, fashioned every aspect of life on Earth in accordance with a spaceship's carrying capacity for astronauts. To them, spaceships were like Noah's Arks, sailing away from an ecologically doomed Earth. They even wrote an homage to Noah as a prologue to their 1974 book, *Ark II:* "Noah had ample warning from a respected authority to build his Ark, and he used his time to good advantage. Skeptics laughed, ridiculed, and drowned—but Noah, the original prophet of doom, survived."[22] "We too have been warned," they continued before plunging into a lengthy analysis of the earth's gloomy biological future and the need to refashion the globe according to the imagined "Ark II" in space.

New soft-tech solutions emerging from space research were presented in the *Whole Earth Catalog,* for example, as attempts to reconcile the natural and technological realms.[23] The editor, Stewart Brand, was a firm believer in the value of space colonization, as he believed that colonies on the moon or Mars could save earthly species from industrial destruction and possible atomic apocalypse. Such colonies could also, at least according to the architect Paolo Soleri, provide humans with a healthy spiritual place.[24] Thus, environmental ethics became a matter of trying to behave like astronauts by adopting "soft" space technologies, such as solar cells. The technology, terminology, and methodology developed for the ecological colonization of space became tools for solving environmental problems on Earth.

Seeing the world as one integrated cabin ecological system was reinforced by the Arab oil embargo of 1973–74, which demonstrated how events on one side of the globe could dominate politics on another. The subsequent literature about alternative energy sources and savings highlighted the importance of seeing local initiatives in a global context. The U.S. National Science Foundation and NASA were suddenly "pouring millions of [research] dollars into solar heating" in an effort to find a viable alternative to oil.[25] The space program would create an entire solar-cell industry. Its clean energy represented for many a possible transition from an age of fossil fuels to a brighter future based on space ecological technologies. The aim of this research was to obtain the same level of self-sufficiency as commandos in harsh territories or astronauts in future space colonies had: "The military has developed a fairly sophisticated technology for the autonomous servicing of permanent and temporary communities in a variety of environments including harsh ones—such as the arctic and space," a commentator noted in *Architectural Design*. "Much of this knowledge could be put to better use. Useful systems include survival packs for pilots bailing out over the arctic [and] life support systems for astronauts."[26]

The ability to see the environment on Earth as a whole, anthropologists and historians have shown, presupposed a privileged point of view from space.[27] This global perspective came with the capability of seeing Earth from outer space. "A view of the earth form outer space gives our generation a perspective never before experienced in history," one urban planner noted, "we are passengers on a planet involved in the intricate cycles of life."[28] The view from outer space triggered planning on a scale that was previously unthinkable. The Greek urban planner and architect Constantinos A. Doxiadis, for example, drew up plans for global settlements. He argued that humanity, thanks to dramatic population growth, was "heading towards a universal city, towards a city which will cover the whole world, towards Ecumenopolis."[29]

IAN MCHARG'S FITTING OF SPACESHIP EARTH

In the area of landscape architecture Ian McHarg became a leading figure in the period. As a professor in the Department of Landscape Architecture at the University of Pennsylvania he inspired a whole generation of landscape architects. In his *Design with Nature* (1969) he based his suggestions for global landscape management on cabin ecological research.

Although McHarg's thinking was considered revolutionary by many of his readers and students, it represented a continuation of an approach to design and architecture that he had learned during his student years at Harvard. Just like Gropius, McHarg would promote science-based modernist architecture and planning, along with respect for nature. For example, McHarg praised the grand landscape design of the Tennessee Valley Authority as "a great vision" because of its commitment to biologically informed planning and restoration.[30] The managerial perspective of large-scale planning, he argued in 1963, was like being "far out in space" looking "back to the distant earth."[31] Environmental problems were the result of the "anarchy which constitutes urban growth" without this proper view from above.[32]

The University of Pennsylvania became the chief employer of former Harvard faculty after Gropius was forced to retire in 1952. His departure led to massive sympathetic resignations of the architecture faculty at Harvard. According to McHarg, G. Holms Perkins, in the Department of Architecture at the University of Pennsylvania, "took advantage of the situation to persuade brilliant faculty, young and old, to join him."[33] Thus, McHarg could engage on a daily basis with former Harvard faculty such as William L. C. Wheaton, Martin Meyerson, Blanche Lemco, and Robert Geddes, as well as Paul Rudolph, who was an annual visitor. Subsequently the department became the leading scene in the United States of modernism in the tradition of Gropius and the Bauhaus, with new faculty members such as Romaldo Giurgola, Robert Venturi, and Louis I. Kahn.

McHarg would frequently refer to the importance of ecology, though it was not until 1966 that this came to the forefront of his approach to landscape architecture. The occasion was the year-long visit to the Department of Landscape Architecture by the South African ecologist John Phillips. Phillips had spent his student years in the same landscape as McHarg, with his mentor Isaac Balfour, the Scottish ecologist and colleague of Patrick Geddes. In South Africa Phillips was known for his close friendship with his patron, the country's longtime prime minister Jan Christian Smuts. Based on Smuts's book *Holism and Evolution* (1926) Phillips developed a holistic theory of ecology whose key concept he termed *biotic community*.[34] Using Smuts's philosophy, Phillips sought to establish a holistic understanding of the environment that explained actions of individual species in the context of the dynamics of an entire biotic community. With this concept he provided Smuts a way to apply ecology to human politics so that Smuts's ideas of racial segregation and managing social communities could have a scientific footing. He endorsed the South African government's design of the Bantustan landscape and settlements through his research on biotic communities. For such beliefs he was eventually expelled from the University of Ghana, and after various consulting jobs for the World Bank, Philips would return to academic life when McHarg invited him to visit the University of Pennsylvania in 1966. Phillips was invigorated by the visit and would later acknowledge the importance of McHarg's work.[35]

It was Phillips who introduced "the holistic approach" to architects and regional planners, arguing that they ought to include ecology and "all forms of life" in their designs.[36] His challenge came in 1968, in a special issue of the journal *Via*, "Ecology in Design," in which a series of planners and designers embraced his views. Jack McCormick wrote that ecological methodology virtually represented "flower power" to planners. Louis Kahn pondered whether to design an ecological garden for his Salk Laboratory (and decided instead to go for a minimalist garden of stones to evoke spiritual "powers of anticipation" among its users). Fritz Morgenthaler and Aldo van Eyck saw in Phillips's holism a return to the wisdom of primitive designs. What all agreed on was the importance of drawing connections between ecology and other disciplines. As Nicholas Muhlenberg noted, "We must consummate a marriage between a reluctant bridegroom (ecologist) and a blushing bride (economist), sending the minister (planner) along on the honeymoon."[37]

McHarg was no less enthusiastic about Phillips and holistic ecology, though Phillips's racist views and support of apartheid neither inspired nor concerned him. What caught McHarg's attention instead was Phillips's holistic approach to landscape architecture. In Phillips's work this approach entailed a total perspective that assigned people to different biotic communities according to their inherent racial qualities and their place on the evolutionary ladder. In McHarg's work ecological management also entailed a holistic perspective assigning people to different environments, but his approach was based not on race but on people's *activities*. In view of destructive industrialism, he saw in his approach an environmental philosophy that enabled humans "to participate in the environment in a way appropriate for survival, and emerge as a fit agent in evolution."[38] McHarg adopted Phillips as his chief mentor. He praised him for his "valuable advice and criticism" in the preface to *Design with Nature,* and later in life he recalled how "the legendary South African ecologist" had contributed "scientific insights" to his book.[39] The book was made possible thanks to a grant from the Conservation Foundation, and he framed the conservation ethic of his patron in Phillips's holistic terms.

The ecological crisis, McHarg argued in *Design with Nature,* had been caused by reckless laissez-faire economy, individualism, Western capitalist greed, chaotic urbanization, fragmentation of social structures, and lack of planning. As a remedy he pointed to the holistic ecology of "the Orient," which was non-anthropocentric and implied orderly planning and respect for the biotic community. Personally, he recalled the Scotland of his childhood, where he could distinguish between the "industrial toil which Glasgow represented and a beautiful countryside" in the city's surroundings.[40] Intellectually, he projected this difference into a grand critique of Western anthropocentric industrialism in contrast to the harmonious naturalism of the Orient. McHarg thus continued the British colonial tradition, embodied in the South African ecologist's thinking, of imagining moral alternatives in the exotic Orient.

In *Design with Nature* the imagined life in outer space represents this holistic "Oriental" alternative to the havoc of Western anthropocentrism. The U.S. space program was well under way; the first unmanned spacecraft landed on the moon while McHarg was writing his book. In the last week of 1968 Apollo 8 sent photographs of Earth as seen from space, one of which McHarg adapted to adorn his book cover. The image of Earth as a whole was to evoke the environmental ethics of the astronaut. "We can use the astronaut as our instructor: he too is pursuing the same quest. His aspiration is survival—but then, so is ours," McHarg argued.[41] The importance of the perspective of the "moon traveler" for understanding ecological relations on Earth had everything to do with the life-support systems of space cabins. The astronaut's photo of Earth as a whole embodied the "Oriental" wisdom of ecological holism, which differed from the destructive compartmental reasoning of the West.

Traveling in space forced the astronaut to recognize humans' biological dependence on the ecological stability of the space cabin. "This realization of dependence was a crushing blow to anthropocentrism," McHarg believed, since the astronaut could not survive if the ship did not sustain its own ecological balance. Earth should be viewed in the same way as the space capsule: "In enlarging the capsule, the objectives remain unchanged: to create a self-sustaining ecosystem—whose only import is sunlight, whose only export is heat—sufficient to sustain a man for a certain period of time." McHarg would emphasize again and again that people on board Spaceship Earth were governed by the same laws as astronauts. The "astronaut's diet," for example, was something the ecologically concerned citizens on Earth should eat, since it was presumably grown within the carrying capacity of a self-sustained space cabin.[42]

In the future, McHarg imagined, humans would build and settle in a "space buoy" located between the moon and the earth. There ecologists would "reproduce a miniature farm" within an artificially built biosphere to provide the astronauts with food. It would be an organic community of plants, insects, fish, animals, and birds with a carrying capacity of several astronauts. The astronaut would function as "a natural scientist and an excellent research ecologist. . . . [His] major task was clearly not only understanding the system, but managing it. Indeed, while the astronaut had learned a great deal of indispensable science, his finest skill was that he could apply this in the management of the ecosystem. We could now call him an intelligent husbandman, a steward."[43] For McHarg, the astronaut and life in the future "space buoy" represented a human ecological utopia.

A future human settlement floating in space became McHarg's conceptual model for landscape architecture on Earth: "The astronaut learned that he had lived in a capsule that was a poor simulation of the earth, but that the world was, indeed, a capsule." Following the strategy of managing energy in the space capsule, he suggested creating "an ecological value system in which the currency is energy."[44] First, an ecosystem inventory should be made of an environment; its changing processes should be investigated and its limiting factors identified. Next, values should be attributed to the ecological aspects of the landscape, permissible and prohibited changes should be determined, and finally indicators of stability and instability in the system should be identified. It was a method designed to determine minimum social cost and maximum social utility for humans as well as nature. Through this utilitarian reasoning, landscape architecture was to mobilize nature's own "intrinsic value-system in which the currency is energy and the inventory is matter."[45]

This value system was based on space ecological analysis, and McHarg used it to analyze landscapes as diverse as the dunes of the New Jersey shoreline, the Richmond Parkway in New York, and the suburban valleys north of Baltimore. Working from the ideal of the perfectly managed closed ecosystem in the imagined space colony, McHarg and his office tried to design these earthly landscapes so that they would gradually turn into environments resembling imagined biotic communities in outer space. McHarg fashioned the landscape architect as a cabin ecological engineer who managed and surveyed the environment in the same way that NASA's ground controllers in Houston kept a close eye on the cabin ecological circulation of energy and materials within a spaceship.

Design with Nature became a phenomenal success, with more than three hundred and fifty thousand copies sold over a period of thirty years. It received several hundred reviews, only one of which, according to McHarg, was critical.[46] The book was taken seriously by scholars, students, administrators, and laypeople alike. Indeed, it changed the field of landscape architecture, which from then on embraced the ideals of space ecosystems. It also changed McHarg's life. He became a celebrity in architecture circles and a kind of cult figure for students, who would gather in large numbers for his lectures. On the occasion of the "Day of Awareness" at the American Institute of Architects' conference in Boston in 1970, for example, McHarg gave an honorary lecture about the importance of ecology for design. The lecture was all about the importance of space-capsule ecology to the study of landscapes: "What's true of the capsule is true of the world," he argued, pointing to the similarities between interacting and recirculation processes in spaceships and what happened in environments on Earth.[47] He said that architects should think of humans as animals in an evolutionary struggle. "We are in this business of adaptation for survival. That is the real definition of architecture," he argued in reference to the Darwinian principle of survival of the fittest. "Architecture should not be called architecture; it should be called fitting."[48]

Humans were not "fitting" very well, McHarg claimed in a series of articles and lectures that became increasingly gloomy during the 1970s. If one looked at human activity from the perspective of a "space voyager," one saw that humans were an "epidemic" and a "disease" destroying the environment at an alarming rate.[49] Humans were not living within the carrying capacity of their closed ecological system. Only through "fitting" of landscapes and buildings could humankind hope to succeed biologically.[50] Architecture should adjust to "basic human needs," a concept based on the astronaut's needs in a space cabin.[51]

FITTING LOCAL SPACE ARKS OF HUMAN SURVIVAL

Ian McHarg's suggestions for remodeling industrially hammered landscapes as space ecological communities were met with widespread support. The activities of the influential New Alchemy Institute may serve as an example as to how some of his ideas were carried out on a local scale.

One of the first designers to build a closed ecosystem was the biologist and New Alchemist John Todd. Space enthusiasts were especially impressed with his experiments with fish farming, which was highly relevant to imagined future farming in outer space. Todd agreed that his experiments had "many of the attributes of a space colony," but in 1977 he would "consider it unsafe to attempt to simulate livable environments [in space] from our present biological knowledge."[52] What he sought to do, in collaboration with a series of ecologists, was build a closed ecosystem on Earth before trying to build one in space. After all, "if stable and productive closed ecosystems could not be made to function on Earth they certainly would not function in orbit," and definitely not on the moon or on Mars.[53] He therefore sought to build closed ecological systems on Earth and develop an ecological managerial system for land and buildings inspired by the ideals of imagined future space colonies.

The New Alchemy projects began in 1969. Trained in agriculture, aquaculture, comparative psychology, and ethnology, Todd was teaching a course in "doom watch biology" at San Diego State University, in California, when he decided to do something about the sad state of the earth.[54] As he later explained to a *New York Times* reporter: "I got tired of ringing the alarm bell all the time. I want constructive alternatives."[55] Together with the oceanographer and fish ecologist William McLarney he founded the New Alchemy Institute to pursue the cause. Their motto, "To Restore the Lands, Protect the Seas, and Inform the Earth's Stewards," captures the spirit of this back-to-the-land commune, which cherished a blend of political anarchism, environmentalism, and anti-urbanism.[56] Scientifically they used ecology and cybernetics in their construction projects, first in 1969 near Woods Hole, Cape Cod, then in 1973 in the province of Limón in Costa Rica, and finally in 1976 on Prince Edward Island, Canada.

The New Alchemists were motivated by a deep-seated fear of not surviving the earth's coming ecological collapse. Their chief metaphorical narrative was the Bible story of Noah, who on God's advice built an ark to save the believers, along with two animals of each species on earth, from the Great Flood. Their entire project revolved around surviving the impending catastrophe, and their strategy was to emulate Noah. They fashioned themselves as "builders of 'lifeboats' and 'arks'" that would be "needed desperately" if humanity was "to avoid famine and hardship" as a result of population growth, rotten capitalism, and greedy exploitation of natural resources.[57] The New Alchemists put their hopes in the construction of a closed ecological lifeboat that would remain afloat even if the larger ecosystem sank. Survival depended on achieving ecological balance and living within nature's carrying capacity, according to an article about their arks in *Science*, since they expected modern agriculture "to collapse, maybe within 10 to 20 years."[58] A *New York Times* reporter visiting the Cape Cod ark in 1976 could not help noticing this "apocalyptic wariness" among the New Alchemists. "Maybe we're only a spark in the dying embers of our civilization," Todd explained.[59]

The New Alchemists' name was inspired by premodern alchemical theories about the reciprocal relationship between microcosms and macrocosms of the world. The ideal house should be like a microcosm of nature's household. As "a productive and self-contained microcosm," the design elements of the arks mirrored the ecological principles of the earth as a whole.[60] Wind generators and greenhouse windows provided the New Alchemy Institute with renewable clean energy, just as the sun provided energy to the earth. A large sun painted on the windmill at Cape Cod was intended to make the point more obvious to visitors, including an engineer who published a report about it in *Science*.[61] Solar-heated fishponds (inspired by fish farming in Maoist China) represented the oceans and provided the residents with fish. In Cape Cod they were covered by a Buckminster Fuller dome that was a mini-representation of the earth. Intensive vegetable gardens mirrored the earth's biota and provided food for the New Alchemists.

The group produced their own power from methane generated by their sewage system, mimicking the earth's chemical processes.[62] An elaborate compost system mimicked the earth's soil processes; it circulated by providing food for a flock of chickens, which represented the earth's birds. Carefully designed buildings integrated the windmill, the fishponds, the gardens, the manure, the composting, the chickens, and the rooms for human activity. As they gained experience with each new ark, the New Alchemists aimed at solar-heated and wind-powered greenhouse-aquaculture buildings. The ark on Prince Edward Island came closest to the ideal. It was built according to their own diagrams for the movement of energy, matter, food, sewage, plants, and humans in the buildings. It did not rely on outside energy and thus came to represent a step toward a self-sufficient architecture that mimicked the ecological processes of nature as a whole.

There were striking similarities between the ark projects of the New Alchemists and ecologically construed space colonies. The former's attempt to escape the environmental destruction on Earth by building arks or spaceships, their progressive idea of being at the forefront of humans' future way of life, their methodological foundation in ecosystem theory, and their belief in the necessity of constructing closed ecological systems for biological survival were based on space ecology. It was "like improving a spaceship while flying through space," commented two visitors who were put to maintain the New Alchemists' ecosystems.[63]

Measured in terms of the number of visitors, the New Alchemy arks were a huge success. By the mid-1970s the ark at Cape Cod had became a "New Age Mecca of sorts," with more visitors than the New Alchemists could handle. Some were put to work on the land, while others were taken on guided tours of the facilities. In effect, the arks evolved into ecotourism resorts, and high admission fees supplied badly needed funding.[64] The message visitors were to take home was that in order to survive the forthcoming ecological catastrophe, it was necessary to build self-sufficient ecological architecture. A journalist visiting in 1976 wrote that Todd pronounced his "evangelical" message "like a high-church Episcopalian," announcing that the New Alchemists had the "means of survival should ecological or economic disaster strike."[65]

Some scientists and architects took great interest in the New Alchemy arks. McLarney involved his friends at the marine biological station in Woods Hole, who in their spare time researched alternative ecological fish-farming technologies.[66] The arks' ability to reduce energy consumption and achieve material self-sufficiency caught the attention of ecologically concerned scholars such as S. David Freeman, Barry Commoner, Herman Daly, Lynn Margulis, and Richard Stein.[67] Architects and designers also were among the visitors. Todd and his wife, Nancy, advised them to build ecological "living machines" (instead of modernist "machines for living"), which were to function as tiny microcosms or mirror images of the macrocosm.[68] Their book *From Eco-Cities to Living Machines: Principles of Ecological Design*, published first in 1980 and revised in 1984 and 1994, was for more than a decade the standard introduction to ecologically informed architecture, complete with advice on how to build with solar panels, bio-toilets, and recirculation of energy and material. The arks came to represent the cutting edge of ecological design.

Despite all these efforts, in the early 1980s the earth was still being "raped biologically" by industrial society, and the need to prepare for the coming ecological doom was as urgent as ever.[69] In order to be fully prepared for the impending catastrophe, the New Alchemists started to experiment with "Ocean Arks," equipped with vegetable greenhouses, freshwater distillation systems, aquaculture pools, animals, and even tree crops (Fig. 7). Following the call of Noah, the ocean arks were designed to save their sailors and species from the coming flood of ecological disasters. They were to be self-sufficient, closed ecological spaceships sailing on the ocean of a dying Spaceship Earth. As one follower argued,

> Noah was told by God to build an ark. . . . There are many signs of the "coming flood" [today]. The overall abuse of the earth by humanity is about to leave our ever growing population "flooded" with survival emergencies, on many levels. . . . Just as Noah needed a life supporting ship that would float independently without access to land, we are in need of life supporting ships that will "float" independently without access to various archaic self-destructive systems which give us acid rain, radioactive waste and power lines lacing the earth like spider webs. . . . We need to evolve self-sufficient living units that *are* their own systems. These units must energize themselves, heat and cool themselves, grow food and deal with their own waste. . . . We are now in need of *Earthships—independent vessels—to sail on the seas of tomorrow.*"[70]

8. THE CLOSED WORLD OF ECOLOGICAL ARCHITECTURE

The ecological colonization of outer and earthly space caused very little controversy until 1975, when royalties from the counterculture *Whole Earth Catalog* were used to finance space research. In the debate that followed among its readers, the overwhelming majority thought that space colonies could provide well-functioning environments for astronauts seeking to push human evolutionary expansion into new territories, while also saving a Noah's Ark of earthly species from industrial destruction and possible atomic apocalypse on Earth. To them space colonies represented a rational, orderly, and wisely managed alternative to irrational, disorderly, and ill-managed Earth. Some of them built Biosphere 2, in Arizona, to prepare for the colonization of Mars and to create an ideal model for how life on Earth should be organized. The skeptical minority argued that space colonization was unrealizable or unethical; nevertheless they adopted the terminology, technology, and methodology of space research in their efforts to reshape the social and ecological matrix on board Spaceship Earth.

THE SPACE-COLONY DEBATE AMONG CO-EVOLUTIONISTS

The financial support of space research from funds generated by the *Whole Earth Catalog* created a fierce debate among its readers, many of whom subscribed and contributed to its sister publication, *Co-Evolution Quarterly*. Stewart Brand, the editor of both publications, would devote much room to the topic in the journal that he later published in book form, *Space Colonies* (1977). The debate represents possibly the first critical reaction of ecologists and environmentalists to the colonization of space.

Gerard O'Neill's articles about the ecological colonization of space were presented in the autumn 1975 issue of *Co-Evolution Quarterly*, with a lengthy interview and a favorable introduction by Brand. Readers were encouraged to voice their opinions about space colonies through a questionnaire and written statements. "Nothing we've run in *The CQ* has brought so much response," Brand noted. Out of 214 replies, 139 (65%) thought the colonization of space was a "good idea," 49 (23%) thought it was a "bad idea," and 26 (12%) were "not sure."[1] The readers of a journal known to be a vanguard for the counterculture, the New Left, and environmentalism thus overwhelmingly supported O'Neill's program.

The large majority saw the colonization of space as worthy of investigation and investment. A leading defender was Buckminster Fuller, who believed that space colonies had been his idea, a claim that was not without merit given the fact that he had published a popular article about ecological cities in outer space in *Playboy* as early as 1968.[2] Although the exobiologist Carl Sagan preferred the term *space cities* to *space colonies*, he too was a supporter of O'Neill's, since his space program would "permit the next evolutionary advance in human society."[3] Tom Heppenheimer argued that engineering agricultural ecological systems in space was not only possible but desirable, since they promised a remedy to environmental ills on Spaceship Earth.[4] A NASA engineer argued that space colonization would "confound" the "limits to growth" thesis advocated by some environmentalists.[5] Environmentalists and ecologists such as David Steindl-Rast, Alan Scrivener, and Carolyn Henson also voiced their support. The popular environmentalist slogan "We have only one Earth," for example, was seen as an example of unsound environmental reasoning.[6] A serious space program, the French oceanographer Jacques Cousteau argued, would provide new technologies for submarine explorations as well as badly needed satellite technology for monitoring the earth.[7] Lynn Margulis, the microbiologist and co-deviser of the Gaia hypothesis, also supported space colonization: "Of course Space Colonies are worthy of investigation and investment," she argued, since human evolution tended to expand into new realms.[8]

The idea of constructing large ecosystems in space met with head-on resistance from the minority of the journal's subscribers. This opposition, which had hardly been voiced until Brand's support of O'Neill, addressed the viability, practicality, and sustainability of building moon bases and exploring deep space with the help of ecological science. Lewis Mumford, for example, saw space colonies as "technological disguises for infantile fantasies."[9] Ken Kesey, the author of *One Flew over the Cuckoo's Nest* (1962), also thought that such a "James Bond" project had "lost its appeal."[10] Likewise, Gary Snyder, the author of *Turtle Island* (1974), thought space colonies were "frivolous."[11] Ernst F. Schumacher, whose *Small Is Beautiful* (1973) had reached a large audience, argued sarcastically that he was "all for it" because space colonies would allow large-scale technocrats to emigrate "out of the way."[12] The advocate of solar energy Wilson Clark did not see a reason to generate solar energy in space, since this could be done more easily on Earth. Dennis Meadows, co-author of *The Limits to Growth* report, also thought the focus should be on solving problems on Earth instead of in outer space. Likewise, Garrett Hardin argued that emigration into space was not a solution for human population growth.[13]

The population biologists Paul and Anne Ehrlich recognized that O'Neill's vision shared "many elements with that of most environmentalists: a high quality of life environment for all peoples, a relatively depopulated Earth in which a vast diversity of other organisms thrive in a non-polluted environment with much wilderness, [and] a wide range of options for individuals." Yet, they argued that space colonization was not a solution to population growth and that biologists "simply have no idea how to create a large stable artificial ecosystem."[14] Environmentalists and ecologists such as Stephanie Mills, Eric Alden Smith, David Brower, Hazel Henderson, and Peter Warshall also voiced their criticism. The biologist and Nobel laureate George Wald viewed space colonies "with horror" as the logical extension of "dehumanization and depersonalization that have already gone much too far on Earth."[15] Most furious perhaps was the poet and environmentalist Wendell Berry, who accused Brand of supporting big government, capitalism, militarism, and "the cult of progress" by devoting *Co-Evolution Quarterly* to space-colony research.[16] He argued that environmentally concerned humanists should abandon the intellectual space capsule that ecologists had created for them. In the lyrics of David Bowie's "Space Oddity," "Now it's time to leave the capsule if you dare."[17]

A third group conditionally supported or criticized O'Neill's proposals. The ecological solar architect Paolo Soleri, for example, was in favor of them but worried that the architectural design of NASA's space colonies failed to address human spiritual needs.[18] The environmentalist William Irwin Thompson pointed out "that the apocalypse that we seek to escape is inside us" and that although there was nothing wrong with setting up a space colony, it failed to nurture "our Buddha-nature."[19] The ecological architect John Todd recognized that his own buildings had "many of the attributes of a space colony," but he still considered it "unsafe to attempt to simulate livable environments [in space] from our present biological knowledge."[20]

The result of these statements was a key consensus paper signed by prominent ecologists and designers, including Ramón Margalef, James Lovelock, Lynn Margulis, and John and Nancy Todd. "The question of space colonization should be explored," they said, but they thought that the colonization of space might lead to unjustifiable exploitation of resources on Earth, and they were unsure about the technological feasibility the project.[21] They believed that a closed ecosystem should be built on Earth before an attempt was made to build one in space. This suggestion became the cornerstone of a series of attempts to create autonomous buildings modeled on spaceships that culminated with one of the most expensive ecological experiments ever, namely, the Biosphere 2 project in Arizona.

THE CAPSULE SYNDROME IN ECOLOGICAL ARCHITECTURE

Designing and building closed autonomous systems became a trend among ecological architects. They struggled to encapsulate buildings so that the inhabitants would be sheltered against the coming doom. Attempts by the cabin ecological industry to transfer its knowledge about space designs to earthly buildings was met with enthusiasm by architects, who responded with proposals for buildings that gradually became more and more self-sufficient and enclosed, reaching a climax in 1991 with the fully encapsulated Biosphere 2.

A leading cabin ecological firm was the Grumman Corporation, which in the 1960s was building planes for the U.S. Air Force and produced aerospace technology for NASA. Grumman played a vital role in the Apollo program by developing and operating the so-called Grumman Lunar Module, in which the corporation took much pride. As the first fully integrated artifact ever designed to operate solely outside the human environment, it was viewed by Grumman employees as a major technological achievement setting the standard for their work. When space business slumped in the early 1970s, Grumman tried to diversify its business by developing products for the civil consumer market. The result was a series of innovative designs that included a modular housing unit based on the Lunar Module, a waste-disposal system inspired by space recirculation technology, a sewage system inspired by the astronauts' toilet, and an energy-efficiency system for homes that incorporated solar cells. These designs and technologies were sold under the label "Grumman's Integrated Household System" and were promoted to architects as an ecological remedy to environmental problems. The system applied technologies and design approaches "initially used in the design of life support systems for spacecraft."[22] Grumman suggested a system of water circulation within the home, for example, that was basically an earthly version of its designs for water circulation and treatment within a spacecraft. Grumman's way of connecting different apparatus within a building into a coherent whole caught the attention of designers. Grumman's study of buildings as a closed ecological system analogous to a closed spaceship raised eyebrows and inspired environmentally concerned architects.

Equally stimulating were new household prototype technologies developed by Lockheed Missiles and Space Company in California. Lockheed argued that technologies it had developed for a lunar base would be ideal for solving many of the ecological imbalances on Earth. Because of its technologies for managing waste, air, food, and energy, the space cabin came to exemplify the rational and scientific way to live ecologically. On board Spaceship Earth, a Lockheed salesperson argued, humans were polluting their cabin with carbon dioxide, they were hardly recycling their waste, and they did not generate enough energy from the sun. It was consequently urgent to transfer technology from space capsules to Earth.[23]

The technology developed by Grumman and Lockheed inspired

projects like the "Integral Urban House" in Berkeley, California. Launched in 1972, it was built as a closed habitat providing an ongoing life-support system for its inhabitants.[24] They read Howard Odum's *Environment, Power, and Society* (1971) and used it to analyze how building designs could contribute to energy management, resource recirculation, and water conservation by regarding a house as a unified whole. By integrating all biotic and a-biotic factors within a closed system, they hoped to construct a building that would function independently as a space cabin. *Life-support system* was a key term borrowed from the space industry that signified a complete system nurturing its inhabitants without relying on resources from the outside world (except rain and energy from the sun).

The Integral Urban House project caught the attention of academics such as Sean Wellesley-Miller and Day Chahroudi, the codirectors of the Solar Energy Laboratory at the Massachusetts Institute of Technology. Impressed by the project and inspired by the New Alchemists, they set out to improve the technical aspect of integral ecological design. The result was the bio-shelter. It was to function as an "autonomous house" with a built-in ecosystem that generated enough food for the residents' basic needs while also providing "shelter" against the immanent ecological collapse of industrial society. It resembled "the ecological bomb shelter" developed by the military, but it was to have a more active community outreach program.[25] The complete self-sufficiency of the bio-shelter was modeled on that of "a space ship," although Wellesley-Miller and Chahroudi did not believe in colonizing outer space.[26]

This type of research was not only a U.S. phenomenon. Alexander Pike and John Frazer, at Cambridge University, formed a similar research group to investigate the relevance of cabin ecological systems to architecture. In response to the worrying news about the ecological state of the world presented at the U.N. Conference on the Environment in Stockholm in 1972, they aimed to construct ecologically autonomous buildings that would function independently of the earth and thus not harm the environment. "We have lost our innocence," Pike argued, referring to the architectural community's support of industrial growth and ecological exploitation. As a remedy he proposed design that aimed at economic "contraction in place of growth, . . . austerity in place of plenty, and . . . the development of a simplified, labour-intensive society to replace the sophisticated, machine-based order that we are now beginning to find so troublesome."[27] Closed ecological buildings promised to be labor intensive and consequently obstacles to economic growth. They should be self-sufficient and thus help to undermine industrial society, while at the same time pointing to a more environmentally friendly future. Despite voicing skepticism toward industrialism and technology, Pike allowed his architecture to be determined by a host of industrial technologies that focused on integrating waste, water, air, and heat technologies into an ecological whole. The aim was to use ambient solar and wind energy, to reduce energy requirements, and to utilize human household and waste material.

One of Pike's students, Brenda Vale, started a "Soft Technology Research Community" to investigate her teacher's thinking on a farm in Montgomeryshire, Wales.[28] The community was to explore inexpensive buildings that did not lose heat and benefited from solar power. Together with her husband, Robert, Brenda Vale built an "autonomous house" that aimed at circulating all its materials and energy on site as a closed ecological system. There was to be no linkage to local water, gas, electricity, or drainage systems. It was to be "a house generating its own power and recycling its own waste."[29] Technically they mobilized hydrogen and oxygen fuel cell units, which had provided auxiliary power in the Gemini and Apollo space capsules. The autonomous house was not a romantic, back-to-the-land vision. Such "dropping out" was "a game for those with private means," the Vales noted in a sarcastic reference to the ultra-hippie Drop City in Colorado. Instead they sought, as the New Alchemists did, to create a shelter in which they could survive the coming doom of industrial society. The construction of autonomous buildings was "important for the survival of mankind" if (or rather when) environmental disaster struck.[30]

In their subsequent work, Brenda and Robert Vale would discuss and evaluate "green" buildings according to their technological and biological performance as closed ecological systems.[31] In these publications, the visual aspect of design was deemed of little importance. At times of deepening environmental crisis, what mattered was to build architecture that could serve as shelter in the coming doom. One important consequence of this ecological approach to design, as the historian of science Michelle Murphy has shown, was a managerial approach to buildings that came to define the sick-building syndrome of the 1990s.³²

Such ideas were also pursued by Kenneth Yeang, a student of Frazer's who matriculated in 1971 and completed a doctoral degree in ecological architecture in 1980. He came to the forefront of ecological architecture during the building boom in Malaysia in the 1980s and early 1990s, though his conceptual thinking dates back to his Cambridge years.³³ At that time Yeang was worried about overpopulation, deterioration of habitats, pollution, radioactive fallout, and suburban sprawl. As a remedy he proposed an ecological approach to architecture through bionics. By imitating processes in nature, architects could find new environmentally friendly designs for human life. The use of biological analogies for design, he argued, would secure "optimum survival" for humans, since such design would benefit from the long evolutionary process of survival of the fittest.³⁴

Yeang's chief example of successful bionics was "a space craft" copying the circulation of matter and energy in nature within a closed artificial ecosystem. According to Yeang, the "space capsule" was like "an autonomous ecosystem" functioning in equilibrium and "completely independent" of its surroundings.³⁵ The spacecraft became the principal module for Yeang's design, which he used to "appraise" a building by making "an orderly inventory of the energy and material inputs and outputs" so that its effect on the environment could be measured.³⁶ Although Yeang saw disadvantages to closing a building to the external environment (except for energy input), he also saw major advantages. The internal circulation of material resources would reduce its environmental impact. Moreover, "by being closed, the internal environment can also be culturally insulated from the cultural context of locality."³⁷ In other words, a building was to be sealed off both environmentally and culturally from industrialism. The creative use of verandah walkways, for example, could allow many buildings, including bioclimatic skyscrapers, or perhaps even cities to be entirely closed off from the external industrial world.

The theoretical underpinnings for these projects came in *Designing with Nature* (1995), a book based on his Cambridge thesis and in its title clearly inspired by McHarg. "In many respects, the problems of survival in an isolated man-made micro-life-support system (as in a spacecraft)," Yeang argued, "resemble the problems encountered in humans' continued survival in the 'global life-support system' or the biosphere."[38] He suggested adapting survival techniques in space by building micro-life-support systems within buildings. His chief source of inspiration in terms of ecology was Odum's analysis of energy flow within space cabins. He borrowed Odum's method of determining the carrying capacity of astronauts in spaceships and used it to determine a building's carrying capacity for inhabitants. He also reused the "life support systems" Odum suggested for astronauts in his technical proposals for "micro-life-support systems" within buildings.[39] Recirculation consequently became a key concept, since ideally buildings would function as spaceships did, receiving only solar energy from their surrounding environment. Yeang would study the role of elevators in order to create more efficient systems for recirculation within a building, stress the importance of optimizing passive modes of energy in closed environments, and investigate whether plants could grow underground as "eco-cells" deep inside developments.[40] Yet measured in terms of enclosure, Young's "eco-cells" were not as radical as the Biosphere 2 project in Arizona.

"NOAH'S ARMY" AT BIOSPHERE 2

The architectural attempt to fully enclose a building ecologically from the surrounding environment came to a climax with the construction of the Biosphere 2 building. Designed by the architect Phil Hawes, it was the first fully enclosed ecosystem and for many architects a model for the future of ecological design. Hawes's chief source of inspiration was the University of Arizona's Environmental Research Lab, which since 1967 had been developing a building that would integrate energy, water, and food into a single ecosystem.[41] Hawes based his drawings on his 1982 outline "Architecture for Space Colonies." It represented a continuation of his projects in New Mexico in the 1970s, which focused on applying space-ecosystem principles to the circulation of energy and materials within a building.[42]

The goals of the privately financed Biosphere 2 project were to make a profit, to prepare for the ecological colonization of space, to build a shelter in which its owners could hide in the event of a serious ecological disaster, and most important, to provide a model for how humans should live within Biosphere 1 (Earth).

The Biosphere 2 idea grew out of discussions at the Synergia Ranch, a commune near Santa Fe, New Mexico. Every hippie commune was unique. This one was based on "discipline and hard work" to carry out projects that could solve the social and ecological crisis on Earth.[43] The hard workers included the union organizer John Allen, Phil Hawes, the philosopher-activist Mark Nelson, and the oil magnate Edward P. Bass. Inspired by ecological futurists such as John McHale and O'Neill, they believed that space technology would play a key role in solving the world's ecological and social problems. Allen and Nelson had cofounded the Institute of Ecotechnics, which aimed at creating synergy between ecological reasoning and technological know-how. They joined Space Biosphere Ventures Inc. to build Biosphere 2.

Bass was the major shareholder in Space Biosphere Ventures, with an investment of $150 million, while his friends from the Synergia Ranch held only symbolic stakes. In view of later criticisms, it is notable that in previous years Bass had sought advice from the Harvard biologist Edward O. Wilson, the Smithsonian biologist Thomas Lovejoy, and the ecological designer Buckminster Fuller, all of whom had advised him to go ahead with the project. His aim, he explained to the *New York Times,* was to profit from the wide public interest in ecology and space colonization. He calculated that Biosphere 2 would turn into a popular ecological "Disneyland" and thus a viable tourist attraction. His partners, such as Allen, Nelson, and Margret Augustine, also regarded Biosphere 2 as a for-profit business project. Their aim was to develop cabin ecological technology for energy efficiency, recycling, waste processing, sewage management, microbial composting, and other emerging solutions to the environmental problems of Spaceship Earth. The hope was that the development and patenting of such technologies would provide Space Biosphere Ventures with a solid profit.[44]

The scientific rationale for Biosphere 2 was to prove that the ecological colonization of space was a viable idea. "Closed ecology systems can free us from Malthusian limitations by making the Solar System our extended home," one proponent argued.[45] Dorion Sagan and Lynn Margulis described the scientific aims in *Biospheres: From Earth to Space* (1989). The book basically applied the Gaia hypothesis to the construction proposal for Biosphere 2 (Fig. 8). "Imagine for a moment you are building a large ship that will travel through space," they encouraged the reader before plunging into a detailed analysis of how the science of ecology could enable people to "live in space indefinitely without the cost of importing supplies." Scientifically it was a question of figuring out how many astronauts an artificial biosphere could support. "Successfully running a new biosphere would show people what it takes to make it in our beloved old one," they argued, pointing to the relevance of such ecological research to "astronauts" on board Spaceship Earth. Moreover, "to settle Mars" with new populated biospheres could provide "protection in case of nuclear war" and "curb global population growth" on Earth.[46] Other ecologists, such as Robert Beyers and Howard Odum, agreed. To them Biosphere 2 was a laboratory and "a prelude to life in space and a means to understand carrying capacity of the earth for humans." Odum, Ramón Margalef, and the founding president of the University Corporation of Atmospheric Research, Walter Orr Roberts, would serve as scientific advisers, making sure Biosphere 2 was built according to cabin ecological design.[47]

Another aim of Biosphere 2 was to provide a shelter in which ecologists and venture partners could survive in co-evolution with thousands of other species if the eco-crisis turned Biosphere 1 into a dead planet like Mars. This sense of a coming doom for Earth prevails in the early literature about the project, which often reminds readers of the story of Noah. "There is an ancient story told in the Bible about a great flood that covered the world long ago," explains a booklet aimed at pupils visiting Biosphere 2. "All of the people and animals were threatened with destruction. But there was one good man named Noah whose family God wanted to save. So He warned Noah that the great flood was coming and told him to build a huge ark." Scientists and designers of Biosphere 2—the "Glass Ark"—fashioned themselves in the image of the biblical Noah. They believed that the new biosphere could secure their personal survival while at the same time saving some of the world's biodiversity. What was needed was a "Noah's army" of environmentalists to protect Spaceship Earth.[48]

Biosphere 2 was completed and sealed in September 1991, after eight "biospherians" dressed in spacesuits marched through the air lock. They promised to stay there for two years. "The project's participants say it can show how to colonize other planets or survive ecological catastrophe on this one," reported a journalist who attended the widely publicized event. The sense of community among old friends from the Synergia Ranch was not appreciated by outside journalists, who described the biospherians as members of a secretive clique. Soon rumors circulated about mechanical (as opposed to organic) rinsing of the ecosystem's carbon dioxide, a bag of supplies smuggled to hungry biospherians, and fresh air being pumped into the building.[49] As a result, a team of scientists lead by Lovejoy and Eugene Odum were called on to scrutinize the science behind the project. With crew members suffering from lack of oxygen, a decision was made to pump more of it into the building, though it effectively ruined the value of the experiment, since the building was supposed to be sealed. Nature did not easily conform to the ecologists' cabin concept, later reviews of the project claimed.[50] The crew members apparently were relieved when they marched out of the air lock—in spacesuits—in September 1993. "The welcoming ceremony, accompanied by a flute solo and a gush of utopian New Age oratory, was in keeping with the odd mix of science and showmanship," the *New York Times* reported on its front page.[51] All of this was unwelcome news to the scientific patrons of the project, such as the Smithsonian Institution's Marine Systems Laboratory, the New York Botanical Garden's Institute of Economic Botany, and the University of Arizona's Environmental Research Laboratory.

The questionable results of the Biosphere 2 experiment led to a dramatic layoff of most of the staff in the spring of 1994. Bass thought "it was time for the project to start making a profit," and he would consequently shift the managerial focus toward ecotourism.[52] More than a half-million visitors had paid $12.95 to learn about the ecological colonization of Mars, and with "biospherians" out of Biosphere 2, ecotourists could now rent rooms within the building and visit a restaurant to experience what ecological life on Mars soon would be like. This was in line with the thinking of ecologists, who believed that ecological microcosms could educate people about ecology, since they could provide pupils with a quick overview of the complexity of nature's economy.[53]

Despite the trouble, ecologists and former biospherians would defend Biosphere 2 and the importance of space colonization for years to come.[54] Scientific experiments, they claimed, were all about learning from mistakes. The project would inspire numerous cabin ecologists working on different schemes to make the Mars environment livable.[55] For architects the Biosphere 2 building became a model for ecological construction, setting the standard for a growing field.[56] Eugene Odum defended Biosphere 2 as a vindication of the Gaia thesis.[57] To Beyers and Howard Odum it proved the viability of the "carrying capacity" concept for the ecological management of Spaceship Earth. Further testing of closed ecosystems in outer space was "long overdue," they argued.[58] In the final scientific report on the Biosphere 2 experiment Howard Odum argued that it had successfully stimulated "the minds of those who have the vision to think beyond the veil of tradition."[59] In fact, Biosphere 2 represented the culmination of research into the ecological colonization of both outer and earthly space.

CONCLUSION: THE UNIFICATION OF ART AND SCIENCE

Ecological designers were concerned about environmental problems in the household of nature as well as in the nature of households. Architectural language traveled into nature and back as the result of mutual stimulation between scientists studying nature and architects designing buildings. This book reviews some of the key moments of inspiration between designers and ecologists, from Bauhaus projects of the interwar period to the eco-arks of the late 1980s.

The unification of art and science is at the heart of the history of ecological design. Bauhaus designers believed that design must follow the laws of nature in order to function effectively. Rightly labeled by one of their contemporaries as "scientific architects," they saw science as a key vehicle for design development.[1] When the Bauhaus faculty settled in London in the mid-1930s in an attempt to reestablish their school, they consequently tried to involve local scientists in their program. Ecologists were among those who responded most favorably.

Former Bauhaus faculty members were key movers in bringing together art and science not only in London but also in the United States in the late 1930s. Walter Gropius, at Harvard, supported nature protection as a response to suburban sprawl and ecological worries. László Moholy-Nagy, at the Chicago Institute of Design, advocated designing objects that would function in harmony with both nature and the human body, while Herbert Bayer, in Aspen, Colorado, developed a visual language of communication that could mediate between a global environmental crisis and individual responsibility.

The Bauhaus designers' approach suggested ways to design with nature according to the functional needs of a complete human being. Their appeal for respecting the boundary between humans and nature was their way of caring for both. Their call for designing according to people's environmental, rational, emotional, and social needs was, in effect, also a plea for greater emphasis on objects and images in historical investigations. Because their environmental designs emphasized imagery, harmony of colors, and proximity between the individual and the world, historical research must also include these areas. The primacy of texts and natural sciences in the hierarchy of today's environmental historiography, for example, may explain why design has been largely ignored by historians of environmentalism and environmental historians alike.

Many ecologists took great interest in design. They saw design as a vehicle for improving human evolutionary fitness and the general state of the world environment. Julian Huxley, for example, actively worked with the designers and used their ideas to envision a new and better way of organizing the British Empire and the earth as a whole. After World War II, ecological engineers took part in technological debates about how to build a closed ecosystem within a space capsule in which humans could survive in outer space; their ideas were instrumental in framing debates about the survival of Spaceship Earth.

The subsequent generation of students and admirers of the Bauhaus, such as Richard Buckminster Fuller and Ian McHarg, would continue the program of trying to unite art and science. It was mainly the new science of space ecological engineering that inspired them. Landscape designers such as McHarg and the New Alchemists took a great interest in these ideas and came to understand their profession as analogous to the management and steering of a spaceship. The imagined life of astronauts became the model for how to live in harmony with nature. The ecologically engineered machinery of the space cabin, such as bio-toilets, solar cells, and recirculation devices, became essential devices for ecological architects such as Fuller, John Todd, and Kenneth Yeang. Closed ecosystems also became models for autonomous buildings that would not harm the environment in which they were placed, such as various bio-arks and the Biosphere 2 project in Arizona.

Thus, space ecology greatly influenced ecological design of the 1960s and beyond. Designing landscapes and buildings in terms of life within space cabins enabled an ecological ethic for humans modeled on the scientifically manageable astronaut. It was an ethic that favored a technological and scientific view of human beings at the expense of wider social and cultural values. The colonization of outer space was of key importance for ecological debate, methodology, and practice. This endeavor grew out of military efforts to improve submarines and shelters and make humans less vulnerable to atomic attack through the dispersion of populations. With the space program of the 1960s, ecological engineers aimed at building cabin ecological systems for astronauts that later served as models for ecological life on Earth. Environmental ethics became a matter of trying to adopt the lifestyle of space travelers, who recirculated their material resources within a closed ecosystem.

Measured in terms of its influence, space ecology was a successful endeavor. Space cabin technologies, such as computer-simulation programs, sewage systems, air-rinsing methodologies, energy-saving devices, and solar cell panels, have become common ecological tools for biological survival. The method of determining a spaceship's "carrying capacity" for astronauts served as a model for organizing humans to live in a practical and ethical way on board Spaceship Earth.

Looking to space ecology as a beacon of hope for an environmentally friendly future had its liabilities. For one thing, it contributed to a culture of scientific technocracy among environmentalists. Moreover, although non-anthropocentric thinkers might question the value of space exploration, they did not question the value of ecological methodology. As a result, ecological analysis has become synonymous with environmental analysis. This ecological colonization of outer and earthly space empowered the managerial ecologist at the expense of humanism. In the process of trying to design for the entire earth, the Bauhaus goal of designing an environment that would function for the entire human being was largely abandoned. The loss of humanism in later ecological designs resulted in buildings and projects detached from their human, social, and natural setting.

A telling example of what ecological architecture came to be in the late 1970s was the attempt in 1976 by architectural students at the University of Minnesota to build their own self-sustaining ecological building with various recirculation devices. They named it Ouroboros "after a mythical dragon which survived by eating its own tail and feces."[2] Like Ouroboros, ecological design in general came to feed on its own ideas and gradually closed itself off from developments in the rest of the design community. The outside world was simply described as "industrial" and thus not worth listening to. As a consequence, many environmentally concerned designers became like astronauts, living intellectually within their own ecological capsules. Indeed, they considered themselves "design outlaws," on the margin of the mainstream, inventing sustainable houses modeled on the household of nature, which would safeguard humans' future ecological existence.[3] Their somewhat narrow focus on the circulation of energy and the efficiency of buildings came at the expense of a wider cultural, aesthetic, and social understanding of architecture and the human condition. As William McDonough and Michael Braungart, two recent environmental architects, have noted about previous ecologically construed buildings, "Efficiency isn't much fun. In a world dominated by efficiency, each development would serve only narrow and practical purposes. Beauty, creativity, fantasy, enjoyment, inspiration, and poetry would fall by the wayside, creating an unappealing world indeed."[4]

While current trends in ecological design are beyond the scope of this book, it is worth noting, however briefly, that when the cold war ended, most environmental designers broke out of the intellectual capsule created for them by ecological space engineering and no longer looked to outer space as a source of inspiration. Those architects concerned with environmentally friendly design who did not endorse space ecology would instead receive attention. One example was Richard Neutra, who considered space research a waste of money.[5] Another was Moshe Safdie, who developed environmentally sensitive and innovative architecture without reference to ecology.[6] Similarly, to avoid harming the landscape, Malcolm Wells chose to build a "gentile architecture" underground that had little to do with space-cabin design principles.[7] As a substitute for space engineering, some designers turned to the perceived ecological wisdom of vernacular architecture and design.[8] With the fall of socialism in the 1990s, others would focus on how ecological building design could also benefit clients financially.[9]

Yet this somewhat narrow neo-liberalist outlook did not move the larger design community to consider the effects its designs might have on the environment. It was not until environmentally concerned citizens and politicians began to demand changes in building techniques that the larger architectural community began to take an interest. One example is Leadership in Energy and Environmental Design (LEED), the rating system for buildings set up by the United States Green Building Council, which encouraged developers of private and public property to reconsider their relationship to both society and nature. Spurred by ecologically minded developers, architects began to explore a host of new, environmentally friendly technologies and building techniques.[10] Rick Cook's designs for the Bank of America building in New York are an example. To meet the new demands, various architectural schools began to provide courses on eco-friendly systems for retrofitting buildings, daylighting, sound water management, and renewable energy.

Trendsetting designers began to take ecological design methods seriously when major clients, contractors, new technologies, and know-how were readily available. For example, Winy Maas and Shigeru Ban brought new and innovative aesthetics to ecological design.[11] Randall Stout, a former project manager for Frank Gehry, is currently designing an "environmental alchemy" in buildings combining both modernist and ecological principles.[12] These architects' aesthetic abilities brought an architectural movement from the margin to the mainstream in architectural journalism. Soon major architectural movers who previously took little interest in ecology began to focus on the issue, as in the Office of Metropolitan Architecture's visions for a city in the desert near Dubai (2006) and Norman Foster's schemes for Masdar City in Abu Dhabi (2007).

This does not mean, of course, that working with nature is about to loose its edge. Key new research groups are promoting the application of ecological methods and systems to computer design and architectural engineering. The Ocean Group network of designers from around the world is currently researching ways in which architects can imitate biological processes in what they call *biomimetics* or *morpho-ecologic* design, while the firm Foreign Office Architects is exploring ways in which sequences of events in the evolutionary development of a species might serve as a principle for architecture in what they call *phylogenesis* in design, while others are investigating *digital-botanic* architecture.[13] There is no doubt about the novelty of these impressive research programs, which in various ways have taken several trendsetting architectural communities and schools by storm.

Yet for all their inventiveness, these efforts are not very different from Moholy-Nagy's bio-technique of the late 1930s. The attempt to base design on biology points to the very core of the modernist heritage reviewed in this book. Indeed, the program of trying to unify art and science may serve as the very definition of the modernist architecture that traces its heritage back to the Bauhaus school. As Julian Huxley once said about the school's founder, "[Gropius's] lifelong aim was to work for the reunification of art and science, without which there can be no true culture."[14]

CAST OF CHARACTERS

Christopher Alexander (1936–) is a British architect who moved to the United States in 1958. He is the coauthor, with Serge I. Chermayeff, of *Community and Privacy: Toward a New Architecture of Humanism* (1963).

Ove N. Arup (1895–1988) was an Anglo-Danish engineer of modernist architecture. He was a structural consultant to Berthold Lubetkin.

Herbert Bayer (1900–1985) was an Austrian graphic designer, architect, and artist who served as director of printing and advertising at the Bauhaus from 1925 to 1928. He moved to New York in 1938 and then to Aspen, Colorado, in 1946. He was the designer of the *World Geo-Graphic Atlas* (1953).

Stewart Brand (1938–) is an American counterculture author. He is the editor of the *Whole Earth Catalog*, first published in 1968, and of the *Co-Evolution Quarterly*.

Marcel L. Breuer (1902–81) was a Hungarian architect and furniture designer. He taught at the Bauhaus, in Germany, in the 1920s, before relocating in 1935 to London and then in 1937 to the United States, where, among other things, he periodically taught architecture at the Harvard School of Design.

Serge Ivan Chermayeff (1900–1996) was a British architect. In 1940 he migrated from London to the United States, where between 1952 and 1970 he taught at Harvard, Yale, and MIT. He was the director of the Institute of Design in Chicago from 1946 to 1951. He was the coauthor, with Christopher Alexander, of *Community and Privacy: Toward a New Architecture of Humanism* (1963).

Wells W. Coates (1895–1958) was an expatriate Canadian architect and cofounder, with Maxwell Fry, of the Modern Architecture Research Group in London in 1933. He designed the Isokon Building, in London, completed in 1934.

Constantinos A. Doxiadis (1913–75) was a Greek architect and town planner known for ekistics, a theory of human settlements. He was the author of *Ecumenopolis* (1963).

Raoul H. Francé (1874–1943) was a Hungarian-born biologist who served as director of the Biological Institute of the German Micrological Society in Munich. He was the author of *Die Pflanze als Erfinder* (Plants as Inventors) (1920).

Edwin Maxwell Fry (1899–1987) was an English architect and cofounder, with Wells Coates, of the Modern Architectural Research Group in London. He practiced with Walter Gropius from 1934 to 1936.

Richard Buckminster Fuller (1895–1983) was an American designer and inventor who developed his own synergetics to create the World Game and design the geodesic dome. He was the author of the *Operating Manual for Spaceship Earth* (1969).

Walter Gropius (1883–1969) was a German architect and founder of the Bauhaus. In 1937 he migrated to the United States, where he headed the Harvard School of Design until 1952.

Julian Sorrell Huxley (1887–1975) was an English evolutionary biologist and ecologist. He was a professor of zoology at King's College, London (1925–27), secretary of the Zoological Society (1935–42), and a member of the think tank Political and Economic Planning. He coauthored, with George P. Wells and Herbert G. Wells, *Science of Life* (1929).

Alexander Korda (1893–1956) was a Hungarian-born film director who worked in London. He directed the science-fiction movie *Things to Come* (1936), based on a novel by H. G. Wells.

James E. Lovelock (1919–) is a British scientist and environmentalist. He was a codeviser, with Lynn Margulis, of the Gaia hypothesis (1974) and is the author of *Gaia: A New Look at Life on Earth* (1979).

Berthold R. Lubetkin (1901–90) was a Russian-born architect who moved to London in 1931, where he founded the Tecton Company. He designed the penguin pool at the London Zoo with the engineering support of Ove Arup.

Ramón Margalef i López (1919–2004) was a Spanish ecologist at the University of Barcelona. He was the author of *Perspectives in Ecological Theory* (1968).

Lynn Margulis (1938–) is an American biologist in the Department of Geosciences, University of Massachusetts, Amherst. She was a codeviser, with James Lovelock, of the Gaia hypothesis of 1974 and is a coauthor, with Dorion Sagan, of *Biospheres from Earth to Space* (1989).

John McHale (1922–78) was a British futurist, artist, and sociologist and worked with Richard Buckminster Fuller at the World Resources Inventory and at the World Design Science Decade Centre, on the Carbondale campus of Southern Illinois University, in the 1960s.

Ian L. McHarg (1920–2001) was a Scottish landscape architect. He moved to the United States after World War II and in 1954 began teaching at the University of Pennsylvania, where he founded the Department of Landscape Architecture in 1957. He was the author of *Design with Nature* (1969).

Peter Chalmers Mitchell (1864–1945) was a British zoologist and secretary of the Zoological Society of London from 1903 to 1935.

László Moholy-Nagy (1895–1946) was a Hungarian constructivist painter and photographer. He was an instructor at the Bauhaus in Germany from 1923 to 1928, then director of the New Bauhaus, in Chicago, for the duration of its existence, from 1937 to 1938. After its closing, he opened the School of Design in 1938 (renamed the Institute of Design in 1944). He was the author of *The New Vision* (1930).

Edward Max Nicholson (1904–2003) was a British ecologist and environmentalist. He was the author of "A National Plan for Great Britain" (1931), an assistant editor of *Week-End Review* (1930–34), and director of the Nature Conservancy (1952–66).

Eugene P. Odum (1913–2002) was an American ecologist known for his work on ecosystems. He founded the Institute of Ecology at the University of Georgia in 1967 and was the author of *Fundamentals of Ecology* (1953).

Howard T. Odum (1924–2002) was an American ecologist known for his work on ecosystems. He was the coauthor, with Elisabeth C. Odum, of *Energy Basis for Man and Nature* (1976).

Gerard Kitchen O'Neill (1927–92) was an American physicist and space activist at Princeton University, where he founded the Space Studies Institute. He was the author of *The High Frontier* (1977).

Walter P. Paepcke (1896–1960) was an American industrialist and philanthropist who started the American Container Corporation in 1926. With the help of Herbert Bayer, he developed Aspen, Colorado, into a ski resort after World War II, and he helped finance the New Bauhaus, in Chicago, in 1938.

John Frederick Vicars Phillips (1899–1987) was a South African ecologist and professor of botany at the University of the Witwatersrand (1931–48). He worked with Ian McHarg and other modernist architects at the University of Pennsylvania in 1966.

Godfrey Samuel (1904–82) was a British modernist architect who worked under Lubetkin in the Tecton architectural firm and was a member of the Modern Architectural Research Group, London. He was the son of the Liberal politician Herbert Samuel.

Paolo Soleri (1919–) is an Italian-American architect who initiated the Acrosanti project in 1970, a planned community combining architecture and ecology.

John Todd (1939–) is an American ecologist. He cofounded, with William McLarney, the New Alchemy Institute in 1969, and he founded Ocean Arks International in 1980. He is the coauthor, with Nancy Todd, of *From Eco-Cities to Living Machines: Principles of Ecological Design* (1994).

Roland Wank (1898–1970) was a Hungarian-born architect who migrated to the United States in 1924. He worked as an architect for the Tennessee Valley Authority from 1933.

Herbert G. Wells (1866–1946) was an English novelist and science-fiction writer. He used human ecology as his chief organizing tool after the publication of *Science of Life* (1929), written with Julian Huxley and George P. Wells.

Bertram Clough Williams-Ellis (1883–1978) was an English-born architect and writer based in Wales who promoted landscape preservation. He was the author of *England and the Octopus* (1928).

Edward O. Wilson (1929–) is an American biologist and ecologist at Harvard University. He is the author of *Sociobiology: The New Synthesis* (1975).

Solly Zuckerman (1904–93), of South Africa, was a zoologist at the Zoological Society, London. He was the author of *The Social Life of Monkeys and Apes* (1932).

NOTES

MANUSCRIPT COLLECTIONS

Archives of University of East Anglia, Norwich, UK.
Herbert Bayer Collection and Archive. Denver Art Museum.
Chermayeff Archive. Avery Library, Columbia University, New York.
Elisabeth H. Paepcke Papers. Special Collections. University of Chicago.
Walter P. Paepcke Papers. Special Collections. University of Chicago.
Registrar Exhibition Files. Museum Archives. Museum of Modern Art, New York.

INTRODUCTION

1. László Moholy-Nagy, "Design Potentials," in *New Architecture and City Planning*, ed. Paul Zucker (New York: Philosophical Library, 1944), 675. For a similar argument, see idem, *Vision in Motion* (Chicago: Paul Theobald, 1947), 44–45. For the motto, see Louis Sullivan, "The Tall Office Building Artistically Considered," *Lippincott's Magazine* 57 (Mar. 1896): 409.

2. Frank Lloyd Wright, *A Testament* (New York: Horizon, 1957), 160, 177.

3. See, for example, Maxwell Fry and Jane Drew, *Tropical Architecture in the Dry and Humid Zones* (London: Batsford, 1964); and Peter Hall, *Cities of Tomorrow: An Intellectual History of Urban Planning and Design in the Twentieth Century* (Oxford: Blackwell, 1996).

4. Nancy Leys Stephan, *Picturing Tropical Nature* (London: Reaktion Books, 2001), 208–45.

5. See Joseph Harris Caton, *The Utopian Vision of Moholy-Nagy* (Ann Arbor, MI: UMI Research Press, 1984); Éva Forgács, *The Bauhaus Idea and Bauhaus Politics,* trans. John Bátki (Budapest: Central European University Press, 1991); Elaine S. Hochman, *Bauhaus: Crucible of Modernism* (New York: Fromm International, 1997); and Margaret Kentgens-Craig, *The Bauhaus and America* (Cambridge, MA: MIT Press, 1999). For a general overview of modernist architecture in Britain, see Henry-Russell Hitchcock, *Modern Archi-*

tecture in England (New York: Museum of Modern Art, 1937), 25–41; F. R. S. Yorke, *The Modern House in England* (London: Architectural Press, 1937); Jeremy Gould, *Modern Houses in Britain, 1919–1939*, Architectural History Monographs, no. 1 (London: Society of Architectural Historians of Britain, 1977); Arts Council of Britain, *Thirties: British Art and Design before the War* (London: Victoria and Albert Museum, 1980); and David Dean, *Architecture of the 1930s: Recalling the English Scene* (New York: Rizzoli, 1983).

6. The exception is an excellent article by Peter Galison, "Aufbau/Bauhaus: Logical Positivism and Architectural Modernism," *Critical Inquiry* 16 (1990): 709–52. See also Judi Loach, "Le Corbusier and the Creative Use of Mathematics," *British Journal for the History of Science* 31 (1998): 185–215; Kenneth Frampton, "The Mutual Limits of Architecture and Science," in *The Architecture of Science*, ed. Peter Galison and Emily Thompson (Cambridge, MA: MIT Press, 1999), 353–73; and Antoine Picon and Alessandra Ponte, eds., *Architecture and the Sciences* (New York: Princeton Architectural Press, 2003).

7. Roderick F. Nash, *The Rights of Nature: A History of Environmental Ethics* (Madison: University of Wisconsin Press, 1989); Holmes Rolston, *Environmental Ethics* (Philadelphia: Temple University Press, 1988); Baird J. Callicott, *In Defense of the Land Ethic* (Albany: State University of New York Press, 1989); Arne Næss, *Ecology, Community and Lifestyle*, ed. and trans. David Rothenberg (Cambridge: Cambridge University Press, 1989); Robyn Eckersley, *Environmentalism and Political Theory: Toward an Ecocentric Approach* (Albany: State University of New York Press, 1992).

8. Donald Worster, *Nature's Economy: A History of Ecological Ideas*, 2nd ed. (Cambridge: Cambridge University Press, 1994).

9. Richard Grove, *Green Imperialism* (Cambridge: Cambridge University Press, 1995); Peder Anker, *Imperial Ecology: Environmental Order in the British Empire, 1895–1945* (Cambridge, MA: Harvard University Press, 2001).

10. Stewart Brand, "The Sky Starts at Your Feet," in *Space Colonies*, ed. Brand (San Francisco: Whole Earth Catalog, 1977), 5.

11. On general design history, see, for example, Gülsüm Baydar Nalbanto-lu and Wong Chong Tahi, eds., *Postcolonial Spaces(s)* (New York: Princeton Architectural Press, 1997); and Edward W. Said, *Orientalism* (New York: Pantheon Books, 1978).

12. Dean Hawkes, *The Environmental Tradition: Studies in the Architecture of Environment* (London: Spon, 1996); Colin Porteous, *The New Eco-Architecture: Alternatives from the Modern Movement* (London: Spon, 2002); Christine Macy and Sarah Bonnemaison, *Architecture and Nature* (New York: Routledge, 2003); John Farmer, *Green Shift: Changing Attitudes in Architecture to the Natural World* (Oxford: Architectural Press, 1999); James Steele, *Ecological Architecture: A Critical History* (London: Thames & Hudson, 2005); David W. Orr, *Design on the Edge* (Cambridge, MA: MIT Press, 2006), 23–39; Pauline Madge, "Design, Ecology, Technology: A Historiographical Review," *Journal of Design History* 6 (1993): 149–65.

13. Pascal Acot, *Historie de l'écologie* (Paris: Presses Universitaires de France, 1988); Stephen Bocking, *Ecologists and Environmental Politics: A History of Contemporary Ecology* (New Haven, CT: Yale University Press, 1997); Frank Benjamin Golley, *A History of the Ecosystem Concept in Ecology* (New Haven, CT: Yale University Press, 1993); Sharon E. Kingsland, "The History of Ecology," *Journal of the History of Biology* 27 (1997): 349–57; Robert P. McIntosh, *The Background of Ecology* (Cambridge: Cambridge University Press, 1985);

Robert E. Kohler, *Landscapes and Labscapes* (Chicago: University of Chicago Press, 2002); Charles T. Rubin, *The Green Crusade* (New York: Free Press, 1994); John Sheail, *Seventy-five Years in Ecology* (Oxford: Blackwell, 1987); Ronald C. Tobey, *Saving the Prairies* (Berkeley and Los Angeles: University of California Press, 1981); Ludwig Trepl, *Geschichte der Ökologie* (Berlin: Athenäum, 1987); Douglas Weiner, *A Little Corner of Freedom* (Berkeley and Los Angeles: University of California Press, 1999); Gregg Mitman, "When Nature *Is* the Zoo: Vision and Power in the Art and Science of Natural History," *Osiris* 2 (1996): 117–43.

14. The exception is a fine article by Sabine Höhler, "'Spaceship Earth': Envisioning Human Habitats in the Environmental Age," *Bulletin of the German Historical Institute* 42 (2008): 65–85. See also Iosef I. Gitelson, G. M. Lisovsky, and Robert D. MacElroy, *Manmade Closed Ecological Systems* (London: Taylor & Francis, 2003), 33–52; Robert J. Beyers and Howard T. Odum, *Ecological Microcosms* (New York: Springer Verlag, 1993), 178–87; Joel B. Hagen, *An Entangled Bank: The Origins of Ecosystem Ecology* (New Brunswick, NJ: Rutgers University Press, 1992), 189–97; Donald A. Beattie, *Taking Science to the Moon* (Baltimore: Johns Hopkins University Press, 2001); Audra J. Wolfe, "Germs in Space," *Isis* 93 (2002): 183–205; Walter A. McDougall, *The Havens and the Earth: A Political History of Space Age* (New York: Basic Books, 1986); James E. Oberg and Alcestis R. Oberg, *Pioneering Space* (New York: McGraw-Hill, 1986), 103–17; Steven J. Dick and James E. Strick, *The Living Universe: NASA and the Development of Astrobiology* (New Brunswick, NJ: Rutgers University Press, 2005); and Steven J. Dick, *The Biological Universe: The Twentieth-Century Extraterrestrial Life Debate and the Limits of Science* (Cambridge: Cambridge University Press, 1996).

15. See Paul N. Edwards, "The World in a Machine: Origins and Impacts of Early Computerized Global Systems Models," in *Systems, Experts, and Computers*, ed. Agatha C. Hughes and Thomas P. Hughes (Cambridge, MA: MIT Press, 2000); Geoffrey C. Bowker, "Biodiversity Datadiversity," *Social Studies of Science* 30 (2000): 643–83; Fernando Elichirigoity, *Planet Management* (Evanston, IL: Northwestern University Press, 1999); and Sharon E. Kingsland, *Modeling Nature*, 2nd ed. (Chicago: University of Chicago Press, 1995).

16. Kevin Kelly, *Out of Control: The Rise of Neo-Biological Civilization* (New York: William Patrick, 1994); Tzvetan Todorov, *Imperfect Garden: The Legacy of Humanism* (Princeton, NJ: Princeton University Press, 2002); Peder Anker, "A Vindication of the Rights of Brutes," *Philosophy and Geography* 7 (2004): 261–66.

17. John D. Bernal, *The Freedom of Necessity* (London: Routledge & Kegan Paul, 1949), 191–213.

18. Galison and Thompson, *Architecture of Science*; Picon and Ponte, *Architecture and the Sciences*; Sylvia Lavin, *Form Follows Libido: Architecture and Richard Neutra in a Psychoanalytic Culture* (Cambridge, MA: MIT Press, 2004); Macy and Bonnemaison, *Architecture and Nature*; Hadas Steiner, "The Forces of Matter," *Journal of Architecture* 10 (2005): 91–109. See also Caroline A. Jones and Peter Galison, eds., *Picturing Science, Producing Art* (New York: Routledge, 1998); and Giuseppa Di Christina, ed., *Architecture and Science* (Chichester, UK: Wiley-Academy, 2001).

19. William Cronon, *Nature's Metropolis* (New York: Norton, 1991); Mike Davis, *Ecology of Fear* (New York: Metropolitan Books, 1998).

20. Kenneth Frampton, *Studies in Tectonic Culture* (Cambridge, MA: MIT Press, 1995), 29–32; Maurice Cranston, *The Noble Savage: Jean-Jacques Rousseau* (Chicago: University of Chicago Press, 1991); Richard Mabey, *Gilbert White* (London: Century, 1986).

21. Shirley Sargent, *John Muir in Yosemite* (Yosemite National Park: Flying Spur, 1971), 18.

22. Peder Anker, "The Philosopher's Cabin and the Household of Nature," *Ethics, Place and Environment* 6 (2003): 131–41; Adam Sharr, *Heidegger's Hut* (Cambridge, MA: MIT Press, 2006).

23. Jack Loeffler, *Adventures with Ed: A Portrait of Abbey* (Albuquerque: University of New Mexico Press, 2002), 151–54.

24. See, for example, Ismo Björn, "Life in the Borderland Forests," in *Encountering the Past in Nature*, ed. Timo Myllyntaus and Mikko Saikku (Athens: Ohio University Press, 2000), 49–73.

25. Percival Goodman, *The Double E* (Garden City, NY: Anchor, 1977); Bruno Latour, *The Politics of Nature: How to Bring the Sciences into Democracy*, trans. Catherine Porter (Cambridge, MA: Harvard University Press, 2004), 131.

CHAPTER 1

1. Patrick Abercrombie, W. G. Constable, Charles Holden, Ian Macalister, and Herbert Read, "Professor Gropius," *Times* (London), 15 Feb. 1937, 8; "Farewell Dinner to Professor Gropius," *Journal of the Royal Institute of British Architects* 44 (20 Mar. 1937): 361, 516. About 140 guests attended the dinner. The invitation is reproduced in David Dean, *Architecture of the 1930s: Recalling the English Scene* (New York: Rizzoli International, 1983), 136–37.

2. Kristina Passuth, *Moholy-Nagy* (London: Thames & Hudson, 1985), 62.

3. László Moholy-Nagy to J. C. Pritchard, 3 Aug. 1935, PP/16/4/1, Archives of University of East Anglia, Norwich, UK. Moholy-Nagy moved out of the building after three months to save money, but he frequently visited colleagues who lived there. Terence Senter, *L. Moholy-Nagy* (London: Arts Council of Great Britain, 1980), 28; idem, "Moholy-Nagy: The Transitional Years," in *Albers and Moholy-Nagy: From the Bauhaus to the New World*, ed. Achim Borchardt-Hume (London: Tate, 2006), 85–91.

4. Isokon Limited, "Schedule of Deeds," PP/16/4/2, Archives of University of East Anglia; Walter Gropius, quoted in Sherban Cantacuzino, *Wells Coates* (London: Gordon Fraser, 1978), 62; Laura Cohn, *Wells Coates* (Oxford: Oxford Polytechnic, 1979); idem, *The Door to a Secret Room* (Aldershot, UK: Scolar, 1999), 41; Anthony Bertram, *The House: A Machine for Living In* (London: A. & C. Black, 1935), 64.

5. Reginald R. Isaacs, *Gropius: An Illustrated Biography of the Creator of the Bauhaus* (Boston: Bulfinch, 1991), 200.

6. Peter Jones, *Ove Arup: Masterbuilder of the Twentieth Century* (New Haven, CT: Yale University Press, 2006), 43–65.

7. Giancarlo Palanti, quoted in Aladar Olgyay, *The Work of Architects Olgyay + Olgyay*, with a letter of introduction by Marcel Breuer (New York: Reinhold, 1952), 15.

8. Aladar Olgyay and Victor Olgyay, *Solar Control and Shading Devices* (Princeton, NJ: Princeton University Press, 1957); Victor Olgyay, *Design with Climate: Bioclimatic Approach to Architectural Regionalism* (Princeton, NJ: Princeton University Press, 1963).

9. Modern Architecture Research Group, *New Architecture: An Exhibition of the Elements of Modern Architecture Organized by the MARS*, exhibition catalog (London: New Burlington Galleries, 1938), 20.

10. Walter Gropius, *The New Architecture and the Bauhaus*, trans. P. Morton Sand (London: Faber & Faber, 1935), 24. I am grateful to Kimberly Collins for this reference.

11. See Lewis Mumford, introduction to Clough Williams-Ellis, *England and the Octopus* (1928; Glasgow: Robert MacLehose, 1975); and David Matless, *Landscape and Englishness* (London: Reaktion Books, 1998), 25–61.

12. Paul Overy, *Light, Air and Openness: Modern Architecture between the Wars* (London: Thames & Hudson, 2007).

13. Clough Williams-Ellis and John Summerson, *Architecture Here and Now* (London: Thomas Nelson, 1934); Clough Williams-Ellis, *Cottage Building* (London: Charles Scribner, 1919); Clough Williams-Ellis and Amabel Williams-Ellis, *The Pleasure of Architecture* (1924; London: Jonathan Cape, 1930). See also Clough Williams-Ellis, ed., *Britain and the Beast* (London: J. M. Dent, 1937); and David Matless, "Appropriate Geography: Patrick Abercrombie and the Energy of the World," *Journal of Design History* 6 (1993): 167–78.

14. Michael Korda, *Charmed Lives* (New York: Random House, 1979), 120–24.

15. Passuth, *Moholy-Nagy*, 67; Reyner Banham, *Theory and Design in the First Machine Age* (New York: Praeger, 1960), 311–19 (also argued in Rainer K. Wick, *Teaching at the Bauhaus* [Ostfildern-Ruit, Germany: Hatje Cantz Verlag, 2000], 131–63). Joseph Harris Caton claims that Moholy-Nagy was unable to recognize the importance of the organic. *Utopian Vision of Moholy-Nagy* (Ann Arbor, MI: UMI Research Press, 1984), 88; see also Victor Margolin, *The Struggle for Utopia* (Chicago: University of Chicago Press, 1997).

16. László Moholy-Nagy, *The New Vision: From Material to Architecture*, trans. Daphne M. Hoffman (New York: Brewster, Warren, & Putnam, 1930). John Summerson, "The Case for a Theory of Modern Architecture," *Journal of the Royal Institute of British Architects* 64 (June 1957): 308. The following discussion is based on the third edition of *The New Version*, published in New York by Wittenborn & Schultz in 1946, and citations are to that edition.

17. Moholy-Nagy, *New Vision*, 29.

18. Jimena Canales and Andrew Herscher, "Criminal Skins: Tattoos and Modern Architecture in the Work of Adolf Loos," *Architectural History* 48 (2005): 235–56; Stephen Jay Gould, *Ontogeny and Phylogeny* (Cambridge, MA: Harvard University Press, 1977); Max Nordau, *Degeneration* (Lincoln: University of Nebraska Press, 1993).

19. Moholy-Nagy, *New Vision*, 16–17.

20. Ibid., 13, 15.

21. Ibid., 16; also argued in L. Moholy-Nagy, *Telehor* (Brno, Czech Republic: Fr. Kalivoda, 1936), 31.

22. Moholy-Nagy, *New Vision*, 18.

23. Raoul H. Francé, *Der Werdegang der Lebenslehre*, quoted in Romain Roth, *Raoul H. Francé and the Doctrine of Life* (Bloomington, IN: 1st Books, 2002), 153–54; idem, *Die Pflanze als Erfinder* (Stuttgart: Kosmos, 1920); idem, *Phoebus: Ein Rückblick auf das glückliche Deutschland im Jahre 1980* (Munich: Drei Masken Verlag, 1927).

24. Siegfried Ebeling, *Der Raum als Membran* (Dessau, Germany: Dunnhapt Verlag, 1926), 12–14; Christoph Asendorf, "Walter Benjamin and the Utopia of the 'New Architecture,'" in *Social Utopias of the Twenties*, ed. Jeannine Fiedler (Wuppertal, Germany: Busmann, 1995), 22–39; Lewis Mumford, *The Culture of Cities* (New York: Harcourt, 1938), 402–93. Mumford's chief source of inspiration was Patrick Geddes; see Volker M. Welter,

Biopolis: Patrick Geddes and the City of Life (Cambridge, MA: MIT Press, 2002). See also Philip Steadman, *The Evolution of Designs* (Cambridge: Cambridge University Press, 1979), 57–73.

25. László Moholy-Nagy, "The New Bauhaus and Space Relationship," *American Architect and Architecture* 151 (Dec. 1937): 24, 28.

26. László Moholy-Nagy, "Modern Art and Architecture," *Journal of the Royal Institute of British Architects* 44 (9 Jan. 1937), 210–13, quotation on 212; idem, "The Concept of Space," in *Bauhaus, 1919–1928,* ed. Herbert Bayer, Walter Gropius, and Ise Gropius (New York: Museum of Modern Art, 1938), 124–26.

27. László Moholy-Nagy, introduction to Walter Gropius, *Rebuilding Our Communities* (Chicago: Paul Theobald, 1945), 11; Walter Gropius, "The Formal and Technical Problems of Modern Architecture and Planning," trans. P. Morton Shand, *Journal of the Royal Institute of British Architects* 41 (19 May 1934): 679–94.

28. Moholy-Nagy, "New Bauhaus and Space Relationship," 28.

29. László Moholy-Nagy, *Painting, Photography, Film,* trans. Janet Seligman (1925; Cambridge, MA: MIT Press, 1967), 13–15.

30. László Moholy-Nagy and Alfred Kemeny, "Dynamic-Constructive Energy Systems" (1922), quoted in *Moholy-Nagy,* ed. Richard Kostelanetz (New York: Praeger, 1970), 29.

31. Moholy-Nagy, *Painting, Photography, Film,* 13–15.

32. László Moholy-Nagy, foreword to *The Street Markets of London,* by Mary Benedetta (London: John Miles, 1936), vii; Bernard Fergusson, *Eton Portrait* (London: John Miles, 1937); John Betjeman, *An Oxford University Chest* (London: John Miles, 1938), 49.

33. Timothy Boon, *Films of Fact: A History of Science in Documentary Films and Television* (London: Wallflower, 2008), 60–62.

34. Moholy-Nagy, "New Bauhaus and Space Relationship," 23.

35. Berthold Lubetkin, quoted in Passuth, *Moholy-Nagy,* 65.

36. Berthold Lubetkin, quoted in John Allan, *Berthold Lubetkin: Architecture and the Tradition of Progress* (London: RIBA, 1992), 199; Oskar Schlemmer, László Moholy-Nagy, and Farkas Molnár, *The Theatre of the Bauhaus,* intro. Walter Gropius, trans. Arthur S. Wensinger (Middletown, CT: Wesleyan University Press, 1961); Adolf Max Vogt, *Le Corbusier, the Noble Savage* (Cambridge, MA: MIT Press, 1998).

37. Allan, *Berthold Lubetkin,* 199–250; idem, *Berthold Lubetkin* (London: Merrell, 2002), 68–71; Peter Guillery, *The Buildings of London Zoo* (London: Royal Commission on the Historical Monuments of England, 1993), 82–85; Peter Coe and Malcolm Reading, *Lubetkin and Tecton: Architecture and Social Commitment* (London: Art Council of Great Britain, 1981), 126–27; Malcolm Reading and Peter Coe, *Lubetkin and Tecton* (London: Triangle Architectural Publications, 1992); Malcolm Reading, "Animals and Men," *Architect's Journal* 195 (5 Feb. 1992): 28–37; Anthony Wylson and Patricia Wylson, *Theme Parks, Leisure Centres, Zoos and Aquaria* (Essex: Longman, 1994), 72; Peter Olney, "The London Zoo and Whipsnade Zoo," in *Great Zoos of the World,* ed. Solly Zuckerman (London: Weidenfeld, 1980), 37–59; Reyner Banham, *Age of the Masters* (New York: Harper & Row, 1962), 80–81.

38. "The Penguin Pool in the London Zoo," *Architectural Review* 76 (July 1934): 17–19. See also George Nelson, "Architects of Europe Today: Tecton," *Pencil Points* 17 (Oct. 1936): 528–40.

39. "The New Penguin Pool in the Zoological Gardens, London," *Architect and Building News* 133 (June 1934): 254–55.

40. *Architects' Journal* 79 (June 1934): 856–59, 873–74, quotation on 857.

41. F. A. Gutheim, "Buildings for Beasts," *Magazine of Art* 29 (Oct. 1936): 455–63, quotations on 462–63.

42. Solly Zuckerman, *From Apes to Warlords: An Autobiography* (London: Hamish Hamilton, 1978), 55–57. The Tecton architects were Berthold Lubetkin, Godfrey Samuel, M. Dugdale, V. Harding, A. Chitty, F. Skinner, and L. Drake.

43. Philip D'Arcy Hurt, *Tuberculosis and Social Conditions in England* (London: National Association for the Prevention of Tuberculosis, 1939).

44. Ove Arup, "Art and Architecture," *Journal of the Royal Institute of British Architects* 73 (Aug. 1966): 350–65; Robert Thorne, "Continuity and Invention," in *Arups on Engineering*, ed. David Dunster (Berlin: Ernst & Sohn, 1996), 234–61.

45. Solly Zuckerman, "The Menstrual Cycle of Primates," *Proceedings of the Zoological Society of London* 100, no. 4 (1930): 691–754; 101, no. 1 (1931): 325–46, 593–602; 102, no. 1 (1932): 139–91; 102, no. 4 (1932): 1059–79.

46. Solly Zuckerman, *The Social Life of Monkeys and Apes* (New York: Harcourt, 1932), 18, 315; Donna Haraway, *Primate Visions* (New York: Routledge, 1989).

47. D. Paul Crook, "Peter Chalmers Mitchell and Antiwar Evolutionism in Britain during the Great War," *Journal of the History of Biology* 22 (1989): 325–56; idem, *Darwinism, War and History* (Cambridge: Cambridge University Press, 1994).

48. Peter Chalmers Mitchell, *My Fill of Days* (London: Faber & Faber, 1936), 362.

49. Ibid., 376.

50. Ibid., 361.

51. "Penguins at the Zoo," *Times* (London), 27 Jan. 1934, 7.

52. "New Gorilla House," ibid., 31 Dec. 1932, 5; "Beauty in New Buildings," ibid., 5 Oct. 1937, 15.

53. L. R. Brightwell, *The Zoo You Knew?* (Oxford: Basil Blackwell, 1936), 221, 230. See also Julian Huxley, *Scientific Research and Social Needs* (London: Watts, 1934), 66; and idem, *Memories* (London: Harper & Row, 1970), 233.

54. Hadas Steiner, "For the Birds," *Grey Room* 13 (2003): 5–31.

55. Peter Chalmers Mitchell, "Logic and Law in Biology," in *Huxley Memorial Lectures* (London: Macmillan, 1932), 1–30.

56. Ronald Aylmer Fisher, *The Genetical Theory of Natural Selection* (Oxford: Clarendon, 1930); Vissiliki Betty Smocovitis, *Unifying Biology: The Evolutionary Synthesis and Evolutionary Biology* (Princeton, NJ: Princeton University Press, 1996); William B. Provine, "The Role of Mathematical Population Geneticists in the Evolutionary Synthesis of the 1930s and 1940s," in *Studies in the History of Biology*, ed. William Coleman and Camille Limoges, vol. 2 (Baltimore: Johns Hopkins University Press, 1978), 167–92.

57. John Burdon Sanderson Haldane and Julian Sorell Huxley, *Animal Biology* (Oxford: Clarendon, 1927).

58. Julian Huxley, "The Biology of Human Nature," *Week-End Review* (London), 28. Jan. 1933, 85–86; 4 Feb. 1933, 114–15; 11 Feb. 1933, 138–39; 18 Feb. 1933, 166–68. Idem, "Diffusion of Culture," ibid., 30 Sept. 1933, 318–19. Ronald Aylmer Fisher and Julian Huxley, "Research in Human Biology," *Times* (London), 14 June 1935, 15.

CHAPTER 2

1. [Edward M. Nicholson], "A National Plan for Great Britain," supplement to *Week-End Review*, 14 Feb. 1931, i–xvi; Edward M. Nicholson, "A Factual Basis for Territorial Planning," *Journal of Town Planning Institute* 22 (1936): 287–91; Reginald R. Isaacs, *Walter Gropius: Der Mensch und sein Werk*, vol. 2 (Berlin: Gebr. Mann Verlag, 1984), 696–97, 788.

2. See, for example, the following by Max Nicholson, "Attack on Poverty," *Week-End Review*, 28 Oct. 1933, 442, 444; "Laissez-Faire and After," ibid., 23 Dec. 1933, 694; and "London Industries," ibid., 16 Sept. 1933, 280, 282.

3. For a larger history of town planning in Britain, see Gordon E. Cherry, *The Evolution of British Town Planning* (New York: Wiley, 1974); idem, *Environmental Planning, 1939–1969*, vol. 2 (London: Her Majesty's Stationery Office, 1975); and Dennis Hardy, *From Garden Cities to New Towns: Campaigning for Town and Country Planning, 1899–1946* (London: Chapman & Hall, 1991).

4. Edward M. Nicholson, "Exhibition," *Week-End Review*, 10 Sept. 1932, 284.

5. Edward M. Nicholson, "An Able and Devastating Document: The Association's Evidence Reviewed," *Town and Country Planning* 6 (1938): 92–93; review of the Royal Commission on the Geographical Distribution of the Industrial Population, in *Minutes of Evidence Taken Before the Royal Commission on the Geographical Distribution of the Industrial Population* (London: His Majesty's Stationery Office, 1937–38); Nicholson, "Attack on Poverty," 442–44; Reginald R. Isaacs, *Gropius: An Illustrated Biography of the Creator of the Bauhaus* (Boston: Bulfinch, 1991), 209.

6. Max Nicholson, "Unemployment Again," *Week-End Review*, 1930, 668–70; idem, "A Yardstick for Industry," ibid., 530–31, quotation on 530.

7. Edward M. Nicholson, "The Political Omnibus," ibid., 1933, 598.

8. See "Uncomfortable England" and "Minimum Standards for Housing," ibid., 16 Dec. 1933, 652–55. A similar manifesto was published in J. H. Badley et al., "Benefits of Light and Air," *Times* (London), 18 Mar. 1932, 10. See also Maxwell Fry, *Autobiographical Sketches* (London: Elek, 1975), 56–159; and Walter Gropius, *Buildings, Plans, Projects, 1906–1969* (Lincoln, MA: I. Gropius, 1972), 17–18.

9. Gregg Mitman, *Reel Nature* (Cambridge, MA: Harvard University Press, 1999), 76–77; Timothy Boon, *Films of Fact: A History of Science in Documentary Films and Television* (London: Wallflower, 2008), 83–89.

10. Julian Huxley, "The Biology of Human Nature," *Week-End Review*, 28 Jan. 1933, 85–86; 4 Feb. 1933, 114–15; 11 Feb. 1933, 138–39; 18 Feb. 1933, 166–68, quotations from introductions to the serial article by Edward M. Nicholson on 85, 114, 138.

11. Julian Huxley, *If I Were Dictator* (London: Methuen, 1934), v–vi. See also Political and Economic Planning Organisation, *The Next Five Years* (London: Macmillan, 1935).

12. Aldous Huxley, *Brave New World* (London: Chatto & Windus, 1932); Ronald W. Clark, *The Huxleys* (New York: McGraw-Hill, 1968), 202; Israel Sieff, *Memoirs* (London: Weidenfeld & Nicolson, 1970), 166.

13. Julian Huxley, *If I Were Dictator*, 24; see also 82–98.

14. Ibid., 80–84 and quotations on 104 and 108.

15. Julian Huxley, "The New Bodleian," *Times* (London), 2 June 1931, 10.

16. Julian Huxley, *Scientific Research and Social Needs* (London: Watts, 1934), 51.

17. Ibid., 56.

18. Julian Huxley, *If I Were Dictator*, 17, 45; Isaacs, *Walter Gropius: Der Mensch und sein Werk*, 2:826–27.

19. Christine Macy and Sarah Bonnemaison, *Architecture and Nature* (London: Routledge, 2003), 137–221. Other studies include Marian Moffett, "Looking to the Future: The Architecture of Roland Wank," *Arris: Journal of the Southeast Chapters of the Society of Architectural Historians* 1 (1989): 5–17; and Marian Moffett and Lawrence Wodehouse, *Building for the People of the United States* (Knoxville: University of Tennessee School of Architecture, 1989).

20. Roland Wank, "The Architecture of Inland Waterways," in *New Architecture and City Planning*, ed. Paul Zucker (New York: Philosophical Library, 1944), 440–58; Lewis Mumford, *The Culture of Cities* (New York: Harcourt, 1938), 344–47; Maxwell Fry, *Fine Building* (London: Faber & Faber, 1944), 143–44.

21. Julian Huxley, "The T.V.A.: A Great American Experiment," *Times* (London), 21, 22 May 1935; idem, *T.V.A.: Adventure in Planning* (Surrey, UK: Architectural Press, 1943); idem, "TVA: An Achievement of Democratic Planning," *Architectural Review* 93 (June 1943): 138–66; Fry, *Fine Building*, 143–44.

22. Julian Huxley, *Memories* (London: Harper & Row, 1970), 233.

23. Julian Huxley, *Scientific Research*, 66.

24. Christopher Frayling, *Things to Come* (London: British Film Institute, 1995); Leon Stover, *The Prophetic Soul* (Jefferson, NC: McFarland, 1987); Karol Kulik, *Alexander Korda* (London: W. H. Allen, 1975), 146–56.

25. See Herbert G. Wells, Julian Huxley, and G. P. Wells, *The Science of Life*, vol. 2 (New York: Doubleday, Doran, 1931), 1027. *The Science of Life* was first published as thirty fortnightly parts from March 1929 till May 1930 by the Amalgamated Press, of London. See also Peder Anker, *Imperial Ecology: Environmental Order in the British Empire, 1895–1945* (Cambridge, MA: Harvard University Press, 2001), 110–16, 196–208; and Piers J. Hale, "Labor and the Human Relationship with Nature," *Journal of the History of Biology* 36 (2003): 249–84.

26. Wells, Huxley, and Wells, *Science of Life*, 1030–32, 1088–90, 1465, 1475–76.

27. Ibid., 1076–88, quotation on 1083.

28. Herbert G. Wells, *The Work, Wealth and Happiness of Mankind*, vol. 1 (New York: Doubleday, Doran, 1931), 35–61, quotation on 35; idem, *The Idea of a World Encyclopedia* (London: Hogarth, 1936), 6.

29. Herbert G. Wells, *The Outline of Human History* (New York: Garden City, 1920), 625, 736.

30. Herbert G. Wells, *What Are We to Do with Our Lives?* (Garden City, NY: Doubleday, 1931), 17.

31. Wells, *Work, Wealth and Happiness*, 232; Le Corbusier, *The Cities of Tomorrow*, trans. Frederick Etchells (London: John Rodker, 1929).

32. "Architectural Changes," *Times* (London), 22 Apr. 1931, 11.

33. Wells, *Work, Wealth and Happiness*, 227.

34. Ibid., 228–300. See also Ronald Aver Duncan, "Science and the Art of Architecture," *Journal of the Royal Institute of British Architects* 37 (7 June 1930): 546–57; and idem, *The Architecture of a New Era* (London: Denis Archer, 1933).

35. Herbert G. Wells, *Experiment in Autobiography* (New York: Macmillan, 1934), 21–25.

36. Herbert G. Wells, *The Shape of Things to Come* (New York: Macmillan, 1933), 389, 408, 425.

37. "The Shape of Things to Come," *Times* (London), 29 Nov. 1933, 12.

38. Herbert G. Wells, "Rules of Thumb for Things to Come," *New York Times*, 12 Apr. 1936; "A Film by Mr. H. G. Wells," *Times* (London), 22 Jan. 1934, 14; "Things to Come," ibid., 29 Oct. 1935, 20; Martin Stockham, *The Korda Collection* (London: Boxtree, 1992), 68; Charles Drazin, *Korda* (London: Sidgwick & Jackson, 2002), 135–43.

39. See Norman Bel Geddes, *Horizons* (Boston: Little, Brown, 1932).

40. Le Corbusier, *Aircraft* (London: Studio, 1935).

41. The dream of being a dictator steering the world based on biological knowledge was not foreign to H. G. Wells; see his book with the telling title *After Democracy* (London: Watts, 1932), 130–33, 192–93.

42. Herbert G. Wells, *Things to Come: A Film* (New York: Macmillan, 1935), 55. This book served as the shooting script for the film that was released in 1936.

43. Sibyl Moholy-Nagy, *Moholy-Nagy: Experiment in Totality* (Cambridge, MA: MIT Press, 1969), 129.

44. Moholy-Nagy's filming techniques and his light-space moderator are described in Eleanor Hight, *Picturing Modernism* (Cambridge, MA: MIT Press, 1995), 65–95.

45. László Moholy-Nagy, "Why Bauhaus Education?" *Shelter*, Mar. 1938, 8–21. The editor of *Shelter* gives Moholy-Nagy credit for the special effects in *Things to Come* on 6–7.

46. Donald Albrecht, *Designing Dreams* (London: Harper & Row, 1986), 162.

47. Vincent Korda, "The Artist and the Film," *Sight and Sound* 3 (1934): 13–16.

48. László Moholy-Nagy, "An Open Letter to the Film Industry," ibid., 56–57. The art directors responsible for camera direction of the film included William C. Menzies, John Bryan, and Fredrick Pusey; see Frayling, *Things to Come*. See also Léon Barsaco, *Caligari's Cabinet and Other Grand Illusions* (Boston: New York Graphic Society, 1976); William Cameron Menzies, "Pictorial Beauty in the Photoplay" (1930), in *Hollywood Directors*, ed. Richard Koszarski (New York: Oxford University Press, 1976), 239–51; Albrecht, *Designing Dreams*, 163; Frayling, *Things to Come*, 16; and Alexander Korda, "British Films: To-Day and To-Morrow," in *Footnotes to the Film*, ed. Charles Davy (New York: Oxford University Press, 1937), 162–69.

49. F. R. S. Yorke and Marcel Breuer, "A Garden City of the Future," *Architects' Journal* 83 (Mar. 1936): 470, 477–82.

50. See Ebenezer Howard, *Garden Cities of Tomorrow* (London: Swan Sonnenschein, 1902), diagram 3; idem, *Garden Cities of Tomorrow*, ed. F. J. Osborn (London: Faber & Faber, 1946), 152n; and Thomas Sharp, *English Panorama* (London: Architectural Press, 1936), 100–101.

51. Wells, *Things to Come: A Film*, 109.

52. Le Corbusier, *Towards a New Architecture*, trans. Frederick Etchells (New York: Payson & Clark, 1927).

53. See Winy Maas, *Metacity-Datatown* (Rotterdam: MVRDV/010, 1999).

54. Wells, *Things to Come: A Film*, 97–98.

55. Ibid., 126, 131.

56. Hardwicke described Wells as "a frustrated pamphleteer who looked more like a prosperous grocer than he bore visible evidence of being one of the true visionaries of our times." Cedric Hardwicke, *A Victorian in Orbit* (Garden City, NY: Doubleday, 1961), 231.

57. Michael Korda, *Charmed Lives* (New York: Random House, 1979), 121.

58. Raymond Massey, *A Hundred Different Lives* (Toronto: McClelland & Stewart, 1979), 191–94, quotations on 191, 192, 194.

59. "Things to Come," *Times* (London), 21 Feb. 1936, 12; Frayling, *Things to Come*, 76–66; Stockham, *Korda Collection*, 71.

60. Thomas Sharp, "The English Tradition in the Town," *Architectural Review* 79 (1936): 163–68.

61. M.S., "Things to Come," ibid., 88–89.

62. "Things to Come," *Architectural Forum* 64 (1936): 420–21.

63. Herbert G. Wells, *The Fate of Homo Sapiens* (London: Secker & Warburg, 1939).

64. Herbert G. Wells, *Guide to the New World* (London: Victor Gollancz, 1941), 92–93, 97.

65. Herbert G. Wells, *All Aboard for Ararat* (New York: Alliance, 1941).

CHAPTER 3

1. John McAndrew, "Bauhaus Exhibition," *Bulletin of the Museum of Modern Art* 6 (Dec. 1938): 5–13; Lorraine Wild, "Europeans in America," in *Graphic Design in America: A Visual History*, by Mildred Friedman et al. (Minneapolis: Walker Art Center, 1989), 153–69; Roland Marchand, *Advertising the American Dream: Making Way for Modernity, 1920–1940* (Berkley and Los Angeles: University of California Press, 1985), 1; Herbert Bayer, Walter Gropius, and Ise Gropius, *Bauhaus, 1919–1928* (New York: Museum of Modern Art, 1938); Herbert Bayer, *Herbert Bayer: Kunst und Design in Amerika, 1938–1985* (Berlin: Bauhaus Archive, 1985).

2. Walter Gropius, "Architecture at Harvard University," *Architectural Record* 81 (May 1937): 8–11, quotation on 11.

3. Walter Gropius, "Essentials for Architectural Education," *PM* 4 (Feb. 1938): 3–16, quotation on 11.

4. Gropius, *Rebuilding Our Communities* (Chicago: Paul Theobald, 1945), 20, 15.

5. Walter Gropius and Martin Wagner, "The New City Pattern for the People and by the People," in *The Problem of the Cities and Towns*, ed. Guy Greer (Cambridge, MA: Harvard University, 1942), 95–116, quotation on 101.

6. Walter Gropius, *Scope of Total Architecture* (1943; New York: Harper & Brothers, 1955), 184–85.

7. Ibid., 184.

8. Maxwell Fry and Jane Drew, *Architecture and the Environment* (London: George Allen & Unwin, 1976). See also C. H. Waddington, *The Man Made Future* (London: Croom Helm, 1978).

9. Adam Rome, *The Bulldozer in the Countryside: Suburban Sprawl and the Rise of American Environmentalism* (Cambridge: Cambridge University Press, 2001).

10. Gropius, *Scope of Total Architecture*, xvii–xviii.

11. Serge Chermayeff and Christopher Alexander, *Community and Privacy: Toward a New Architecture of Humanism* (Garden City, NY: Doubleday, 1963). See also *Design and*

the Public Good: Selected Writings, 1930–1980, by Serge Chermayeff, ed. Richard Plunz (Cambridge, MA: MIT Press, 1982), 156–57; and Alan Powers, *Serge Chermayeff* (London: RIBA, 2001), 37–61.

12. Chermayeff and Alexander, *Community and Privacy*, 46–47. On the importance of suburban sprawl to the environmental debate, see Rome, *Bulldozer in the Countryside*.

13. Walter Gropius to Serge Chermayeff, 28 Sept. 1966, Chermayeff Archive, Avery Library, Columbia University, New York.

14. "Ecological Architecture: Planning the Organic Environment," *Progressive Architecture* 47 (May 1966): 120–34, quotation on 121.

15. Robert Sommer, "The Ecology of Privacy," *Library Quarterly* 36 (1966): 234–48.

16. Ian L. McHarg, *Conversations with Students* (New York: Princeton Architectural Press, 2007), 64; idem, *A Quest for Life: An Autobiography* (New York: Wiley, 1996), 65–92.

17. James Sloan Allen, *The Romance of Commerce and Culture: Capitalism, Modernism, and the Chicago-Aspen Crusade for Cultural Reform*, rev. ed. (Boulder: University Press of Colorado, 2002). See also Bernhard Widder, *Herbert Bayer* (Vienna: Springer Verlag, 2000), 57–79.

18. László Moholy-Nagy, introduction to Gropius, *Rebuilding Our Communities*, 11.

19. Ibid., 12. For an excellent discussion of Moholy-Nagy and Bauhaus design in Chicago, see Allen, *Romance of Commerce and Culture*.

20. László Moholy-Nagy, *Vision in Motion* (Chicago: Paul Theobald, 1947), 5, emphasis in original.

21. László Moholy-Nagy, "Why Bauhaus Education?" *Shelter*, Mar. 1938, 8–21, quotations on 15, 12.

22. Ibid., 20–21.

23. Ibid., 12.

24. László Moholy-Nagy, "The Concept of Space," in Bayer, Gropius, and Gropius, *Bauhaus, 1919–1928*, 124; Herbert Bayer, "Fundamentals of Exhibition Design," *PM* 6 (Dec. 1939): 17–25.

25. Robert Snyder, *The World of Buckminster Fuller*, VHS (New York: Mystic Fire Video, 1971).

26. See the Navy Register of Commissioned and Warrant Officers, at the U.S. Naval Academy Archive, Annapolis, MD. I am grateful to Gary A. LaValley, at the U.S. Naval Academy Archive, for information about Fuller's status in the navy; e-mail messages to author, 11 and 20 Feb. 2003. Biographies include Athena Lord, *Pilot for Spaceship Earth* (New York: Macmillan, 1978), 27–33; and Lloyd Steven Seiden, *Buckminster Fuller's Universe: An Appreciation* (New York: Plenum, 1989), 52–66. Fuller is not mentioned in the Admiral's autobiography; see Albert Gleaves, *The Admiral* (1934; Pasadena, CA: Hope, 1985). See also Peter Karsten, *The Naval Aristocracy: The Golden Age of Annapolis and the Emergence of Modern American Navalism* (New York: Free Press, 1972).

27. Richard Buckminster Fuller, quoted in Lord, *Pilot for Spaceship Earth*, 33. See also Fuller's *Utopia or Oblivion: The Prospects for Humanity* (New York: Overlook, 1969), 1–12.

28. See Joel Mokyr, *The Lever of Riches: Technological Creativity and Economic Progress* (Oxford: Oxford University Press, 1990); Alexander Morris Carr-Saunders, *The Population Problem: A Study in Human Evolution* (Oxford: Clarendon, 1922); and John A. Garraty, *The Great Depression* (San Diego: Harcourt Brace Jovanovich, 1986).

29. See John M. Staudenmaier, "Two Technocrats, Two Rouges: Henry Ford and Diego Rivera and Contrasting Artists," *Polhem: Tidskrift För Teknikhistoria* 10 (1992): 2–28; and Charles S. Maier, "Society as Factory—Between Taylorism and Technocracy," in *In Search of Stability: Explorations in Historical Political Economy* (Cambridge: Cambridge University Press, 1987), 19–69.

30. Richard Buckminster Fuller, *4D Time Lock* (1928; Albuquerque, NM: Biotechnic, 1972), 1–2.

31. Ibid., 6–7; Richard Buckminster Fuller, "Tree-like Style of Swelling Is Planned," *Chicago Evening Post*, 18 Dec. 1928, 5.

32. See "Dymaxion," *Harvard Crimson*, 22 May 1929.

33. Frank Lloyd Wright, "Ideas for the Future," *Saturday Review*, 17 Sept. 1938, 14–15; Joseph Connors, "Wright on Nature and the Machine," in *The Nature of Frank Lloyd Wright*, ed. Carol Bolon, Linda Seidel, and Robert Nelson (Chicago: University of Chicago Press, 1988), 1–19. On Fuller's relationship with other architects, see Felicity D. Scott, "On Architecture under Capitalism," *Grey Room* 6 (2002): 44–65.

34. Inez Cunningham, "Fuller's Dymaxion House on Display," *Chicago Evening Post*, 13 May 1930.

35. Richard Buckminster Fuller, "Putting the House in Order," *Shelter*, Nov. 1932, 2–9, quotation on 2; Robert Ezra Park and Ernst W. Burgess, *Introduction to the Science of Sociology* (Chicago: University of Chicago Press, 1921); Roderick Duncan McKenzie, "The Ecological Approach to the Study of the Human Community," in *The City*, ed. Robert Ezra Park, Ernest W. Burgess, and Roderick Duncan McKenzie (Chicago: University of Chicago Press, 1925), 63–73; idem, "The Scope of Human Ecology," *Papers and Proceedings: The American Sociological Society* 20 (1925): 141–54.

36. Fuller, "Putting the House in Order," 3, 4, 6. See also D. Paul Crook, *Darwinism, War and History: The Debate over the Biology of War from the 'Origin of Species' to the First World War* (Cambridge: Cambridge University Press, 1994).

37. Richard Buckminster Fuller, *Nine Chains to the Moon* (1938; Garden City, NY: Anchor, 1971), 14, emphasis in original.

38. Ibid., 18–29.

39. Ibid., 41. See also Richard Buckminster Fuller, "Universal Architecture," *T-Square*, Feb. 1932, 2–16.

40. Fuller, *Nine Chains to the Moon*, 38.

41. [Richard Buckminster Fuller], "U.S. Industrialization," *Fortune*, Feb. 1940, 50–57. See also "New Era," *Time*, 5 Feb. 1940; and "After Five Years of Dymaxion History," *Fortune*, July 1932, 22–26.

42. Fuller, *Nine Chains to the Moon*, 106.

43. Alexander Dorner, *The Way beyond Art: The Work of Herbert Bayer* (New York: Wittenborn, 1947), 131; also argued in Jan van der Marck, *Herbert Bayer* (Boston: Nimrod, 1977), 5. For a full discussion of Bayer's life, see the excellent biography by Gwen Finkel Chanzit, *Herbert Bayer and Modernist Design in America* (Ann Arbor, MI: UMI Research Press, 1987), reprinted as *From Bauhaus to Aspen: Herbert Bayer and Modernist Design in America* (Boulder, CO: Johnson Books, 2005); and idem, *Herbert Bayer* (Seattle: University of Washington Press, 1988).

44. Carl E. Schorske, *Fin-de-Siècle Vienna: Culture and Politics* (New York: Knopf,

1980); Allan Janik, *Wittgenstein's Vienna Revisited* (New Brunswick, NJ: Transaction Books, 2001); Sylvia Lavin, *Form Follows Libido: Architecture and Richard Neutra in a Psychoanalytic Culture* (Cambridge, MA: MIT Press, 2004).

45. Éva Forgács, *The Bauhaus Idea and Bauhaus Politics,* trans. John Bátki (Budapest: Central European University Press, 1991); Elaine S. Hochman, *Bauhaus: Crucible of Modernism* (New York: Fromm International, 1997); Margaret Kentgens-Craig, *The Bauhaus and America* (Cambridge, MA: MIT Press, 1999); Rainer K. Wick, *Teaching at the Bauhaus* (Ostfildern-Ruit, Germany: Hatje Cantz Verlag, 2000).

46. Clark V. Poling, *Kandinsky: Russian and Bauhaus Years* (New York: Solomon R. Guggenheim Museum, 1983), 36–56; Herbert Bayer, *Herbert Bayer: Das künstlerische Werk, 1918–1938,* trans. George L. Mosse (Berlin: Mann, 1982).

47. Herbert Bayer, quoted in L. Sandusky, "The Bauhaus Tradition and the New Typography," *PM* 4 (June 1938): 24. See also idem, "Towards a Universal Type," ibid. 6 (Dec. 1939): 27–32.

48. Herbert Bayer, foreword to *Herbert Bayer: The Complete Work,* by Arthur A. Cohen (Cambridge, MA: MIT Press, 1984), xi; Peter Galison, "Aufbau/Bauhaus: Logical Positivism and Architectural Modernism," *Critical Inquiry* 16 (1990): 709–52; Kenneth Frampton, "The Mutual Limits of Architecture and Science," in *The Architecture of Science,* ed. Peter Galison and Emily Thompson (Cambridge, MA: MIT Press, 1999), 353–73.

49. Joella Bayer to Elisabeth Paepcke, 5 Jan. 1973, box 23, folder 7, Elisabeth H. Paepcke Papers, Special Collections, University of Chicago. See also Jean-Jacques Rousseau, *Discourse on the Origin on Inequality,* trans. Donald A. Cress (New York: Hackett, 1992).

50. Herbert Bayer, "Contribution toward Rules of Advertising Design," *PM* 6 (Dec. 1939): 7; Percy Seitlin, "Herbert Bayer," ibid., 1, 26, 32; Herbert Bayer, *Herbert Bayer: Photographic Works* (Los Angeles: Center for Visual Arts, 1977); Ulrich Pohlmann, "El Lissitzky's Exhibition Designs," in *El Lissitzky: Beyond the Abstract Cabinet,* by Margarita Tupitsyn, with contributions by Ulrich Pohlmann and Matthew Drutt (New Haven, CT: Yale University Press, 1999), 52–64.

51. Herbert Bayer, "International Design Conference," *Print* 9 (July–Aug. 1955): 12.

52. José Luis Sert, *Can Our Cities Survive? An ABC of Urban Problems, Their Analysis, and Their Solutions* (Cambridge, MA: Harvard University Press, 1942).

53. Herbert Bayer, "Notes of Exhibition Design," *Interiors* 106 (July 1947): 60–77; Doris Brian, "Bayer Designs of Living," *Art News* 42 (Mar. 1943): 20; Mary Anne Staniszewski, *The Power of Display: A History of Exhibition Installations at the Museum of Modern Art* (Cambridge, MA: MIT Press, 1998), 143–235.

54. "Maps and Global Cartography," *Life,* 3 Aug. 1942, 57–65, Registrar Exhibition Files, box 236, Museum Archives, Museum of Modern Art, New York (hereafter MoMA).

55. Richard Edes Harrison, *Look at the World: The Fortune Atlas for World Strategy* (New York: Knopf, 1944); editors of *Fortune,* "The Logic of the Air," in *Compass of the World,* ed. Hans W. Weigert and Vilhjalmur Stefansson (New York: Macmillan, 1945), 121–36; Irving Fisher, "A World Map," *Geographical Review* 33 (1943): 605–19.

56. "No Spot on Earth is More Than 60 Hours From Your Local Airport," advertisement by Consolidated Aircraft Corporation, *Newsweek,* 8 Mar. 1943, 3; American Airlines, "War-Thinking," *New York Herald Tribune,* 9 Feb. 1943, 84, MoMA, box 236.

57. Monroe Wheeler to L. F. V. Drake, 28 May 1943, MoMA, box 236.

58. Wheeler to Roy Stryker, Office of War Information, 15 Mar. 1943, ibid.

59. Some of the illustrations and panels were used in "Sky-Roads," a traveling exhibition organized by the Civil Aeronautics Administration in collaboration with the Museum of Modern Art, later published as "Weather and Warfare," *Skyways*, Feb. 1944, 41–44, 47–50. Bayer based some of his artwork on Consolidated Vultee Aircraft Corporation, *Maps and How to Understand Them* (New York, 1943). See also Robert Marc Friedman, *Appropriating the Weather: Vilhelm Bjerknes and the Construction of a Modern Meteorology* (Ithaca, NY: Cornell University Press, 1993).

60. On the history of panoramas, see Stephan Oettermann, *The Panorama: History of a Mass Medium* (New York: Zone Books, 1997).

61. Wendell L. Willkie, "Airways to Peace," *Bulletin of the Museum of Modern Art* 11 (Aug. 1943): 3–21; idem, *One World* (New York: Simon & Schuster, 1943).

62. R. Roger Remington and Robert S. P. Fripp, *Design and Science: The Life and Work of Will Burtin* (Burlington, VT: Lund Humphries, 2007), 71–78.

63. Wheeler to John R. Fleming, 7 Jan. 1943, MoMA, box 236.

CHAPTER 4

1. John Beardsley, *Earthworks and Beyond* (New York: Abbeville, 1989), 87; David Bourdon, *Designing the Earth* (New York: Abrams, 1995), 210; Suzaan Boettger, *Earthwork: Art and Landscape of the Sixties* (Berkeley and Los Angeles: University of California Press, 2002), 175; Sue Spaid, *Ecovention: Current Art to Transform Ecologies* (Cincinnati: The Contemporary Art Center, 2002), 10; Eric Higgs, *Nature by Design: People, Natural Process, and Ecological Restoration* (Cambridge, MA: MIT Press, 2003); David W. Orr, *The Nature of Design: Ecology, Culture and Human Intention* (Oxford: Oxford University Press, 2002).

2. Herbert Bayer, *World Geo-Graphic Atlas: A Composite of Man's Environment* (Chicago: Container Corporation of America, 1953); Philip B. Meggs, *A History of Graphic Design*, 2nd ed. (New York: Van Nostrand Reinhold, 1992), 325. Historians and sociologists of cartography who do not discuss Bayer include Mark Monmonier, in *Mapping It Out* (Chicago: University of Chicago Press, 1993); Denis Cosgrove, in *Apollo's Eye: A Cartographic Genealogy of the Earth in the Western Imagination* (Baltimore: Johns Hopkins University Press, 2001) and "Maps, Mapping, Modernity: Art and Cartography in the Twentieth Century," *Imago Mundi* 57 (2005): 35–54; Norman J. W. Thrower, in *Maps and Civilization* (Chicago: University of Chicago Press, 1996); Daniel Dorling and David Fairbairn, in *Mapping: Ways of Representing the World* (Edinburgh Gate, Harlow, UK: Longman, 1997); and Karen Piper, in *Cartographic Fictions: Maps, Race, and Identity* (New Brunswick, NJ: Rutgers University Press, 2002).

3. See John B. Harley, "Introduction: Text and Contexts in the Interpretation of Early Maps," in *From Sea Charts to Satellite Images: Interpreting North American History through Maps*, ed. David Buisseret (Chicago: University of Chicago Press, 1990), 3–15; idem, "Maps, Knowledge, and Power," in *The Iconography of Landscape: Essays on the Symbolic Representation, Design and Use of Past Environments*, ed. Denis Cosgrove and Stephen Daniels (Cambridge: Cambridge University Press, 1988), 277–312; Denis Wood, *The Power of Maps* (New York: Guilford, 1992), 191; Margaret Beck Pritchard and Henry G. Talia-

ferro, *Degrees of Latitude: Mapping Colonial America* (Williamsburg, VA: Abrams, 2002); James Corner, "The Agency in Mapping," in *Mappings*, ed. Denis Cosgrove (London: Reaktion Books, 1999), 213–52; and Geoff King, *Mapping Reality: An Exploration of Cultural Cartographies* (London: Macmillan, 1996).

4. Walter P. Paepcke, "The 'Great Ideas' Campaign," *Advertising Review* 2 (Fall 1954): 25–28, quotation on 28; Georgine Oeri, "Great Ideas of Western Man," *Graphis* 13 (1957): 504–13; Daniel J. Boorstin, *The Image* (New York: Atheneum, 1978), 189; Roland Marchand, *Creating the Corporate Soul: The Rise of Public Relations and Corporate Imagery in American Big Business* (Berkeley and Los Angeles: University of California Press, 1998), 335.

5. Daniel Catton Rich, "Modern Art in Advertising," introduction to *Modern Art in Advertising: Designs for the Container Corporation of America,* by Paul Theobald (Chicago: Container Corporation of America, 1946), 7; W.A.H., "World-Famed Artists in the Service of Advertising," *Graphis* 1–2 (Sept–Oct. 1945): 82–87, 104; Neil Harris, "Designs on Demand: Art and the Modern Corporation," in *Art, Design, and the Modern Corporation,* by Neil Harris and Martina Roudabush Norelli (Washington, DC: Smithsonian Institution Press for the National Museum of American Art, 1985), 8–30.

6. See Sigfried Giedion, "Herbert Bayer and Advertising in the U.S.A.," *Graphis* 1 (Nov–Dec. 1945): 348–58, 422.

7. "Prepackaged War," *Fortune*, Dec. 1941, 86–89, 168, quotation on 86; Susan Black, ed., *The First Fifty Years: 1926–1976* (Chicago: Container Corporation of America, 1976), 31; Alexander Weaver, *Paper, Wasps, and Packages* (Chicago: Container Corporation of America, 1937); Samuel P. Hays, *Conservation and the Gospel of Efficiency: The Progressive Conservation Movement, 1890–1920* (Cambridge, MA: Harvard University Press, 1959).

8. Walter Paepcke to Elisabeth Paepcke, telegram, 2 Mar. 1945, and Herbert Bayer to Walter Paepcke, 5 Mar. 1945, box 96, file 9, Walter P. Paepcke Papers, Special Collections, University of Chicago (hereafter Walter Paepcke Papers).

9. Walter Paepcke to Herbert Bayer and Joella Bayer, 14 June, 1945, ibid.

10. Walter Paepcke to Herbert Bayer, 22 May 1945, ibid.; Paepcke says much the same thing in Walter Paepcke to Bayer, 31 May 1945, ibid.

11. Walter Paepcke to Joella Bayer, 1 Nov. 1945, ibid.

12. Herbert Bayer, "A Statement for an Individual Way of Life," *Print* 16 (May–June 1962): 26–33, quotation on 26.

13. Joella Bayer to Walter Paepcke, 14 Feb. 1946, box 96, file 10, Walter Paepcke Papers; *Oral History of Serge Chermayeff,* interview by Betty J. Blum (Chicago: Art Institute of Chicago, 1986), 95.

14. James Sloan Allen, *The Romance of Commerce and Culture: Capitalism, Modernism, and the Chicago-Aspen Crusade for Cultural Reform,* rev. ed. (Boulder: University Press of Colorado, 2002), 73–76; Finis Dunaway, *Natural Visions: The Power of Images in American Environmental Reform* (Chicago: University of Chicago Press, 2005).

15. "Personality in Print: Herbert Bayer," *Print* 9 (July–Aug. 1955): 33–43, quotation on 35.

16. Bayer, "Statement for an Individual Way of Life," 28. An expression of similar ideas by Bayer is quoted in *Great Ideas,* ed. John Massey (Chicago: Container Corporation of America, 1976), xi.

17. Walter Paepcke to V. E. Ringle, *Aspen Times*, 2 Apr. 1946; Charles C. Eldredge, foreword to Harris and Norelli, *Art, Design, and the Modern Corporation,* 6.

18. Walter Paepcke to Herbert Bayer, telegram, 3 Oct. 1946, box 96, file 10, Walter Paepcke Papers.

19. Dean Sims, "The Town That Came Back for Management," *Manage,* Nov. 1956, 22–25; Leo Lionni, *Between Worlds* (New York: Knopf, 1997), 159–63; Bayer, *World Geo-Graphic Atlas,* 148.

20. Herbert Bayer, *Painter, Designer, Architect* (New York: Reinhold, 1967), 113.

21. Herbert Bayer to Walter Paepcke, 14 Apr. 1946, box 96, file 10, Walter Paepcke Papers; Jan van der Marck, *Herbert Bayer* (Boston: Press, 1977), 7.

22. Bayer, *Painter, Designer, Architect,* 150.

23. On agency in nature, see, for example, Michel Serres, *The Natural Contract,* trans. Elizabeth MacArthur and William Paulson (Ann Arbor: University of Michigan Press, 1996); Bruno Latour, *The Pasteurization of France,* trans. Alan Sheridan and John Law (Cambridge, MA: Harvard University Press, 1988); and Donna Haraway, *The Companion Species Manifesto* (Chicago: Prickly Paradigm, 2003).

24. George Renner, quoted in Herbert Bayer, "Why Container Corporation Publishes an Atlas," typescript, Herbert Bayer Collection and Archive, Denver Art Museum.

25. Bayer, preface to *World Geo-Graphic Atlas,* 4.

26. Walter Paepcke to Joella Bayer, 17 Apr. 1953, box 96, file 13, Walter Paepcke Papers.

27. Bayer, "Statement for an Individual Way of Life," 32; Walter P. Paepcke, "Why Container Corporation Publishes an Atlas," in Bayer, *World Geo-Graphic Atlas,* 5.

28. Herbert Bayer, "Goethe and the Contemporary Artist," *College Art Journal* 11 (1951): 37–40; Martin J. S. Rudwick, "The Emergence of a Visual Language for Geological Science, 1760–1840," *History of Science* 14 (1976): 149–95; Robert J. Richards, *The Romantic Conception of Life* (Chicago: University of Chicago Press, 2002).

29. Bayer, *World Geo-Graphic Atlas,* 27, 71, 204.

30. Ibid., 225 and quotation on 277.

31. Herbert Bayer, "On Trademarks," in *Seven Designers Look at Trademark Design,* ed. Egbert Jacobsen (Chicago: Paul Theobald, 1952), 48–52, 49.

32. Herbert Bayer, "Toward the Book of the Future," in *Books for Our Time,* ed. Marshall Lee (New York: Oxford University Press, 1951), 22–25, quotation on 25; idem, introduction to *Exhibitions and Displays,* by Erberto Carboni (Milan: Silbana, 1957), 5–11.

33. Herbert Bayer, "Notes on world geo-graphic atlas," MS, Herbert Bayer Collection and Archive, quotations on 4, 5; idem, *World Geo-Graphic Atlas,* 4.

34. Herbert Bayer, "International Design Conference," *Print* 9 (July–Aug. 1955): 12; Bayer expressed similar ideas in idem, "Notes on world geo-graphic atlas," 5. See also Powers, *Serge Chermayeff,* 175.

35. Bayer, "Goethe and the Contemporary Artist," 38; László Moholy-Nagy and Alfred Kemeny, "Dynamic-Constructive Energy Systems" (1922), in *Moholy-Nagy,* ed. Richard Kostelanetz (New York: Praeger, 1970), 29; László Moholy-Nagy, *Painting, Photography, Film,* trans. Janet Seligman (1925; Cambridge, MA: MIT Press, 1967), 13–15.

36. "Directory of Owners of the Color Harmony Manual," 31 Aug. 1951, and Walter Paepcke to Robert L. Stearns, 19 Sept. 1950, box 30, files 8 and 7, respectively, Walter Paepcke Papers.

37. Serge Chermayeff to Walter Paepcke, 8 Jan. 1959, box 8, file 9, ibid. A parallel story about patronage, environmentalism, and design involved Chermayeff and Walter Paepcke; see box 2, Chermayeff Archive.

38. Egbert Jacobsen, *Basic Color: An Interpretation of the Ostwald Color System* (Chicago: Paul Theobald, 1948), 54, inspired by Wilhelm Ostwald, *Die Farbenfibel* (Leipzig: Verlag Unesma, 1916); Egbert Jacobsen, *The Color Harmony Manual* (Chicago: Container Corporation of America, 1942); "Prepackaged War," 86–89; John Gage, *Colour and Meaning: Art, Science and Symbolism* (London: Thames & Hudson, 1999), 241–68; Philip Ball, *Bright Earth: The Invention of Colour* (London: Viking, 2001), 356–59.

39. Bayer, "Notes on world geo-graphic atlas," quotation on 3.

40. Sibyl Moholy-Nagy, review of *World Geo-Graphic Atlas*, by Herbert Bayer, *College Art Journal* 14 (Winter 1955): 177–78.

41. Bayer, *World Geo-Graphic Atlas*, 191.

42. Ibid., 4.

43. Herbert Bayer, "My Position as a Non Scientist," *Print* 9 (July–Aug. 1955): 44.

44. Ellsworth Huntington, *Principles of Human Geography* (New York: Wiley, 1949); Bayer, *World Geo-Graphic Atlas*, e.g., 190.

45. Bayer, *World Geo-Graphic Atlas*, 278.

46. Walter Paepcke to Russell Lynes, *Harper's Magazine*, 11 Apr. 1959; Walter Paepcke to Cass Canfield, Harper & Brothers, 4 June 1959; and Herbert Bayer to Sandy Doughtly, Houghton Mifflin, 13 Nov. 1959, all in box 25, file 3, Walter Paepcke Papers.

47. See Stevenson's personal message to Paepcke in Adlai E. Stevenson, *The Stark Reality of Responsibility* (Chicago: Americana House, 1952), copy in box 181, file 1, ibid.; and Porter McKeever, *Adlai Stevenson* (New York: William Morrow, 1989).

48. Edward Weeks, The Atlantic Bookshelf, *Atlantic Monthly*, Feb. 1954, 76, 78; Robert E. Fulton, review of *World Geo-Graphic Atlas*, by Herbert Bayer, *Saturday Review*, 20 Mar. 1954, 37.

49. Reviews of Bayer's *World Geo-Graphic Atlas* by Edward L. Ullman in *Geographical Review* 45 (Jan. 1955): 147–49 and by H. Täubert in *Petermanns Mitteilungen* 98 (1954): 230.

50. J.F.M., "Bayer's Geo-graphics," *Industrial Design* 1 (1954): 94–97; Groff Conklin, "World Geo-Graphic Atlas," *Print* 9 (July–Aug. 1955): 44–51, quotation on 46.

51. Edward Imhof, review of *World Geo-Graphic Atlas*, by Herbert Bayer, *Graphis* 11 (1955): 428–33.

52. Sibyl Moholy-Nagy, review of Bayer, *World Geo-Graphic Atlas*, 178.

53. See, for example, William Van Royen, *Atlas of the World's Resources: The Agricultural Resources of the World* (New York: Prentice-Hall, 1954); idem, *Atlas of the World's Resources: The Mineral Resources of the World* (New York: Prentice-Hall, 1952); and Rand McNally, *International World Atlas* (New York, 1961).

54. See Tony Loftas, ed., *Atlas of the Earth*, with a foreword by Sir Julian Huxley (London: Mitchell Beazley, 1971); A. L. Farley, *Atlas of British Columbia: People, Environment, and Resource Use* (Vancouver: University of British Columbia Press, 1979); and Cartography Department of the Clarendon Press, *Oxford Economic Atlas of the World*, 2nd ed. (Oxford: Oxford University Press, 1972).

55. Rand McNally, *The Earth and Man World Atlas*, with a foreword by Julian Huxley (New York, 1972).

56. See Kenneth MacLean and Norman Thomson, eds., *Problems of Our Planet: An Atlas of Earth and Man* (Edinburgh: Bartholomew, 1977); Rand McNally, *Our Magnificent Earth: Atlas of Earth Resources* (New York, 1979); and Geoffrey Lean and Don Hinrichsen, *Atlas of the Environment* (Oxford: Helicon, 1990).

57. Ben Crow and Alan Thomas, *Third World Atlas* (Milton Keynes, PA: Open University Press, 1983, 1984, 1985, 1986, 1988).

58. See Norman Myers, ed., *The Gaia Atlas of Planet Management,* 2nd ed. (New York: Doubleday, 1993); idem, ed., *Gaia: An Atlas of Planet Management* (Garden City, NY: Anchor, 1984); idem, *The Gaia Atlas of Future Worlds* (New York: Doubleday, 1990); Lee Durrell, *Gaia: State of the Ark Atlas* (New York: Doubleday, 1986); Michael Kidron and Ronald Segal, *The New State of the World Atlas* (New York: Simon & Schuster, 1981), with revised editions in 1984, 1987, and 1991; Joni Seager, ed., *The State of the Earth Atlas* (New York: Simon & Schuster, 1990); Julian Burger, *The Gaia Atlas of First Peoples* (London: Robertson McCarta, 1990); and Herbert Girardet, *The Gaia Atlas of Cities* (London: Gaia Books, 1992). See also Brian Groombridge and Martin D. Jenkins, *World Atlas of Biodiversity* (Berkeley and Los Angeles: University of California Press, 2002).

59. Jeremy Black, *Maps and Politics* (Chicago: University of Chicago Press, 1997), 64, 82.

60. Penny Jones and Jerry Powell, "Garry Anderson Has Been Found!" *Resource Recycling,* May 1999, 25–26.

CHAPTER 5

1. John Cage, "Diary: How to Improve the World (You will Only make Matters Worse)," *New Literary History* 3 (1971): 201–14, quotation on 210.

2. Christine Macy and Sarah Bonnemaison, *Architecture and Nature* (New York: Routledge, 2003), 317–18.

3. Daniel Treiber, *Norman Foster* (London: Spon, 1995), 38–39; Norman Foster, "Richard Buckminster Fuller," in *Buckminster Fuller: Anthology for the New Millennium,* ed. Thomas T. K. Zung (New York: St. Martin's, 2001), 1–8; idem, *Rebuilding the Reichstag* (London: Weidenfeld, 2000), 138–39.

4. A complete list of hundreds of recent and old books, films, links, and articles by and about Fuller is available at www.bfi.org. One Web site, www.cjfearnley.com/buckyrefs .html, lists no fewer than 181 links to Fuller-related material on the Web. Notable publications include Richard Buckminster Fuller, *Cosmography* (New York: Macmillan, 1992); idem, *Your Private Sky* (Baden, Switzerland: Lars Müller, 1999); idem, *Your Private Sky: Discourse* (Baden: Lars Müller, 2001); idem, *Buckminster Fuller, Inventions: Twelve Around One,* exhibition catalog (Cincinnati: Carl Solway Gallery, 1981); Lloyd Steven Seiden, *Buckminster Fuller's Universe: An Appreciation* (New York: Plenum, 1989); Martin Pawley, *Buckminster Fuller* (London: Trefoil, 1990); J. Baldwin, *Bucky Works: Buckminster Fuller's Ideas for Today* (New York: Wiley, 1996); Zung, *Buckminster Fuller;* and Chris Zelov and Phil Cousineau, eds., *Design Outlaws on the Ecological Frontier* (Easton, PA: Knossus, 1997).

5. James C. Scott, *Seeing Like a State* (New Haven, CT: Yale University Press, 1998), 103–46. See also David Milne, "The Artist as a Political Hero," *Political Theory* 8 (1980): 525–45. On self-fashioning, see Mario Biagioli, *Galileo Courtier* (Chicago: University of Chicago Press, 1993).

6. Richard Buckminster Fuller, *Utopia or Oblivion: The Prospects for Humanity* (New York: Overlook, 1969), 54–79, 80–113; Athena Lord, *Pilot for Spaceship Earth* (New York: Macmillan, 1978), 101–2.

7. "R. Buckminster Fuller's Dymaxion World," *Life,* 1 Mar. 1943, 40–55. The various maps are reprinted in Fuller, *Your Private Sky,* 250–75. My reading is inspired by Denis Wood,

The Power of Maps (New York: Guilford, 1992). The map received a favorable review from the cartographer Irving Fisher in "A World Map," *Geographical Review* 33 (1943): 605–19.

8. See Susan Schulten, *The Geographical Imagination in America, 1880–1950* (Chicago: University of Chicago Press, 2001), 204–38; and Alan K. Henrikson, "The Map as an 'Idea': The Role of Cartographic Imagery during the Second World War," *American Cartographer* 2, no. 1 (1975): 19–53.

9. Peter Pearce, *Structure in Nature Is a Strategy for Design* (Cambridge, MA: MIT Press, 1978).

10. "Marines Try Out Flyable Shelter," *New York Times*, 29 Jan. 1954; Robert W. Marks, *The Dymaxion World of Buckminster Fuller* (Carbondale: Southern Illinois University Press, 1960), 133, 198–201, 202–9; John McHale, *R. Buckminster Fuller* (New York: George Braziller, 1962), 32; Tony Robbin, *Engineering a New Architecture* (New Haven, CT: Yale University Press, 1996), 38.

11. Richard Buckminster Fuller, "New Directions," *Perspecta*, Summer 1952, 29–37, quotations on 33.

12. Alex Soojung-Kim Pang, "Dome Days: Buckminster Fuller in the Cold War," in *Cultural Babbage*, ed. Francis Spufford and Jennifer Uglow (Boston: Faber & Faber, 1997), 167–92.

13. Richard Buckminster Fuller, "Why Not Roofs over Our Cities?" *Think Magazine*, Jan.–Feb. 1968, 8–11.

14. Richard Buckminster Fuller, *Ideas and Integrities: A Spontaneous Autobiographical Disclosure*, ed. Robert W. Marks (Englewood Cliffs, NJ: Prentice-Hall, 1963), 269; idem, *Critical Path* (New York: St. Martin's, 1981), xxiv.

15. Richard Buckminster Fuller to Don Metz, Yale School of Architecture, 2 Oct. 1964, Chermayeff Archive.

16. Richard Buckminster Fuller, *Basic Biography* (Philadelphia, 1975).

17. "Modern Living," *Time*, 18 Jan. 1964, 46–51, quotations on 46.

18. Marks, *Dymaxion World of Buckminster Fuller*; McHale, *R. Buckminster Fuller*; Thomas H. Garver, *Two Urbanists* (Waltham, MA: Rose Art Museum, Brandeis University, 1965).

19. Fuller, *Ideas and Integrities*, 187, emphasis in original.

20. Ibid., 244.

21. Ibid., 93, 113.

22. Richard Buckminster Fuller, *Operating Manual for Spaceship Earth* (Edwardsville: Southern Illinois University Press, 1969), 46, emphasis in original.

23. Ibid., 44, 133.

24. Fuller, *Ideas and Integrities*, 256–63; idem, *Education Automation* (1962; London: Jonathan Cape, 1973), 76; idem, *World Design Science Decade, 1965–1975*, ed. John McHale, 2 vols. (Carbondale: Southern Illinois University Press, 1963–67).

25. Fuller, *Utopia or Oblivion*, 293; idem, *Education Automation*, 62–65.

26. Fuller, *Ideas and Integrities*, 257.

27. Richard Buckminster Fuller, "Vision 65 Summary Lecture," 23 Oct. 1965, Southern Illinois University, published in Fuller, *Utopia or Oblivion*, 117.

28. Several historians of early modern science have argued that challenges to social order often translated into threats to nature itself. See, for example, Simon Schaffer, "The Earth's Fertility as a Social Fact in Early Modern Britain," in *Nature and Society in*

Historical Context, ed. Mikuláš Teich, Roy Porter, and Bo Gustafsson (Cambridge: Cambridge University Press, 1997); and Lorraine Daston and Katharine Park, *Wonders and the Order of Nature* (New York: Zone Books, 1998). I believe this argument can be extended to modern society as well.

29. Richard Buckminster Fuller, "Planned Implementation of the World Resources Simulation Center," 18 June 1969, MS, AA 737 F96 F9633, Avery Library, Columbia University, quotation on 5.

30. Herbert G. Wells, *World Brain* (London: Methuen, 1938).

31. Thomas B. Turner, "World Game State-of-the-Art Report," Dec. 1969, MS, AA 737 F96 F9633, Avery Library, Columbia University, quotation on 4.

32. Richard Buckminster Fuller, "World Game: How It Came About," in *50 Years of the Design Science Revolution and the World Game* (Carbondale: World Resources Inventory, Southern Illinois University, 1969), 111–18, quotation on 114.

33. Richard Buckminster Fuller, Edwin Schlossberg, and Daniel Gildesgame, *World Game Report* (New York: New York Studio School of Painting and Sculpture, 1969), 11.

34. Fuller, *50 Years of the Design Science Revolution*, v.

35. Richard Buckminster Fuller, testimony on 4 Mar. 1969 before the Senate Subcommittee on Intergovernmental Relations, S. Res. 78, 91st Cong., 1st sess., p. 3; John von Neumann and Oskar Morgenstern, *Theory of Games and Economic Behavior* (Princeton, NJ: Princeton University Press, 1944).

36. Richard Buckminster Fuller, "The World Game—How to Make the World Work" (1967), in *Utopia or Oblivion*, 157–61.

37. Fuller, *Utopia or Oblivion*, 206.

38. Fuller, Schlossberg, and Gildesgame, *World Game Report*, 1, 2, 14.

39. Medard Gabel, *Energy, Earth, and Everyone: A Global Energy Strategy for Spaceship Earth* (San Francisco: Straight Arrow Books, 1975), 155.

40. Fuller, Schlossberg, and Gildesgame, *World Game Report*, 1; Fuller, *Operating Manual for Spaceship Earth*, 131; Paul R. Ehrlich, *The Population Bomb* (New York: Ballantine Books, 1968).

41. See Donella H. Meadows et al., *The Limits to Growth* (New York: Universe Books, 1972); Paul N. Edwards, "The World in a Machine: Origins and Impacts of Early Computerized Global Systems Models," in *Systems, Experts, and Computers*, ed. Agatha C. Hughes and Thomas P. Hughes (Cambridge, MA: MIT Press, 2000), 201–53; and Fernando Elichirigoity, *Planet Management* (Evanston, IL: Northwestern University Press, 1999). For a broader review of the scientific debates, see Chunglin Kwa, "Representations of Nature Mediating between Ecology and Science Policy," *Social Studies of Science* 17 (1989): 413–42; and Sharon E. Kingsland, *The Evolution of American Ecology* (Baltimore: Johns Hopkins University Press, 2005), 206–31.

42. John Markoff, *What the Dormouse Said: How the 60s Counterculture Shaped the Personal Computer Industry* (New York: Viking, 2005).

43. Bill Voyd, "Funk Architecture," in *Shelter and Society*, ed. Paul Oliver (New York: Praeger, 1969), 156–64, quotation on 156; Robert Snyder, *World of Buckminster Fuller*, VHS (New York: Mystic Fire Video, 1971); Stewart Brand, ed., *Whole Earth Catalog* (Menlo Park, CA: Portola Institute, 1969), with new editions in 1971, 1980, 1994.

44. See Richard Buckminster Fuller with Jerome Agel and Quentian Fiore, *I Seem to Be a Verb* (New York: Bantam Books, 1970); Richard Buckminster Fuller, *Buckminster*

Fuller to Children of Earth (Garden City, NY: Doubleday, 1972); and idem, *Tetrascroll* (New York: St. Martin's, 1975).

45. See Hugh Kenner, *Bucky: A Guided Tour of Buckminster Fuller* (New York: William Morrow, 1973), 163–99; Douglas Davis, *Art and the Future* (New York: Praeger, 1973), 184–85; Alden Hatch, *Buckminster Fuller: At Home in the Universe* (New York: Crown, 1974); Donald W. Robertson, *Mind's Eye of Richard Buckminster Fuller* (New York: Vantage, 1974); Lord, *Pilot for Spaceship Earth; Buckminster Fuller: Autobiographical Monologue/Scenario,* ed. Robert Snyder (New York: St. Martin's, 1980); Richard Buckminster Fuller, "Technology and the Human Environment," in *The Futurists,* ed. Alvin Toffler (New York: Random House, 1972), 298–306; and Macy and Bonnemaison, *Architecture and Nature,* 293–340.

46. James Ridgeway, *The Politics of Ecology* (New York: E. P. Dutton, 1970), 14.

47. Garrett Hardin, *Exploring New Ethics for Survival: The Voyage of the Spaceship Beagle* (New York: Viking, 1972); idem, "Living on a Lifeboat," *BioScience* 20 (Oct. 1974): 561–68. See also Kenneth E. Boulding, "The Economics of the Coming Spaceship Earth," in *Environmental Quality in a Growing Economy,* ed. Henry Jarrett (Baltimore: Johns Hopkins Press, 1966), 3–14. A typical critique of Fuller's technological optimism is found in William Ophuls, *Ecology and the Politics of Scarcity* (San Francisco: Freeman, 1977), 158–59.

48. Richard Buckminster Fuller, *Earth, Inc.* (Garden City, NY: Anchor, 1973), 142.

49. Fuller, "World Game," 116; Meadows et al., *Limits to Growth.*

50. Richard Buckminster Fuller, "City of the Future," *Playboy,* Jan. 1968, 166–68, quotation on 167. See also E. J. Appelwhite, *Synergetics Dictionary: The Mind of Buckminster Fuller,* 4 vols. (New York: Garland, 1986).

51. Richard Buckminster Fuller, "The Year 2000," *Architectural Design,* Feb. 1967, 92–95.

52. Fuller, "City of the Future," 166–68.

53. Fuller, *Utopia or Oblivion,* 155–56.

54. Richard Buckminster Fuller, "Education for Comprehensivity," in *Approaching the Benign Environment,* by Richard Buckminster Fuller, Eric A. Walker, and James R. Killian (London: Collier Books, 1970), 15–111, quotation on 110.

55. Richard Buckminster Fuller, *Synergetics: Explorations in the Geometry of Thinking* (New York: Macmillan, 1975). The technical studies include Edward Popko, *Geodesics* (Detroit: School of Architecture, University of Detroit, 1968); J. C. Bohlen, *Trigonometric Relations for Geodesic Domes* (Vancouver: Department of the Environment, 1974); Hugh Kenner, *Geodesic Math* (Berkeley and Los Angeles: University of California Press, 1976); and Magnus J. Wenninger, *Spherical Models* (Cambridge: Cambridge University Press, 1979). The more practical studies include Pacific Domes, *Domebook 2* (Bolinas, CA, 1972); John Prenis, ed., *The Dome Builder's Handbook* (Philadelphia: Running Press, 1973); Peter Hjersman, *Dome Notes* (Berkeley, CA: Erewon, 1975); The Big Outdoors People, *The Dome Plan Book* (Wyoming, MN, 1978); William Yarnall, *Dome Builder's Handbook No. 2* (Philadelphia: Running Press, 1978); and Gene Hopster, *How to Design and Build Your Dome Home* (Tucson, AZ: H. P. Books, 1981).

56. Victor Papanek, *Design for the Real World: Human Ecology and Social Change* (New York: Pantheon Books, 1971), xxvi.

57. Fuller, *Critical Path,* xvii; Gabel, *Energy, Earth, and Everyone,* 8.

58. Richard Buckminster Fuller, *Grunch of Giants* (New York: St. Martin's, 1983), 1–2.

59. Fuller, *Your Private Sky: Discourse*, 177–225.
60. K. Michael Hays and Dana Miller, eds., *Buckminster Fuller: Starting with the Universe* (New York: Whitney Museum of American Art, 2008).

CHAPTER 6

1. Richard Dawkins, *The Extended Phenotype: The Gene as the Unit of Selection* (Oxford: Oxford University Press, 1982), 162, 235.
2. Edward O. Wilson, *Sociobiology: The New Synthesis* (Cambridge, MA: Harvard University Press, 1975), 548.
3. Garland B. Whisenhunt, "A Life Support System for a Near Earth or Circumlunar Space Vehicle," *Astronautical Sciences Review*, July–Sept. 1960, 13–20, quotation on 14; Jack Myers, "Introductory Remarks," *American Biology Teacher* 25 (1963): 409–11. For a technical review of space ecology, see Eugene B. Koonecci, "Space Ecological Systems," in *Bioastronautics*, ed. Karl E. Schaefer (New York: Macmillan, 1964), 274–304; and James Stephen Hanrahan and David Bushnell, *Space Biology: The Human Factors in Space Flight* (London: Thames & Hudson, 1960).
4. W. B. Cassidy, preface to *Bioengineering and Cabin Ecology*, ed. Cassidy (Tarzana, AZ: American Association for the Advancement of Science, 1969), ix; C. C. Brock, "Space Flight under Sea," quoted in "Space Medicine," by Hubertus Strughold, *Missiles and Rockets* 3 (Feb. 1958): 186. See also "Sub Environment Studied as Space Prototype," ibid. 4 (Sept. 1958): 19.
5. Office of Civil Defense, *Shelter Design Data*, vol. 1, *Shelter Design and Analysis* (Washington, DC: Department of Defense, 1967), pt. 8, p. 1. See also the following by the OCD: *Fallout Shelter Surveys: Guide for Architects and Engineers* (Washington, DC: Department of Defense, 1960); *The Family Fallout Shelter* (Washington, DC: Department of Defense, 1962); *Emergency Operating Centers* (Washington, DC: Department of Defense, 1964); *Family Shelter Designs* (Washington, DC: Department of Defense, 1966); and *Shelter through Architectural Design* (Washington, DC: Department of Defense, 1967). In addition, see Roger S. Cannell, *Live: A Handbook of Survival in Nuclear Attack* (Englewood Cliffs: Prentice-Hall, 1962); and Arthur I. Waskow and Stanley L. Newman, *America in Hiding* (New York: Ballantine Books, 1962), 5.
6. Tom Vanderbilt, *Survival City: Adventures among the Ruins of Atomic America* (New York: Princeton Architectural Press, 2002), 34–36.
7. Ralph E. Lapp, *Must We Hide?* (Cambridge, MA: Addison-Wesley, 1949), 165. See also Sylvan G. Kindall, *Total Atomic Defense* (New York: Richard R. Smith, 1952), 84–85; and Peter Galison, "War against the Center," in *Architecture and the Sciences*, ed. Antoine Picon and Alessandra Ponte (New York: Princeton Architectural Press, 2003), 196–227.
8. Frieda B. Taub, "Some Ecological Aspects of Space Biology," *American Biology Teacher* 25 (1963): 412–21; idem, "Closed Ecological Systems," *Annual Review of Ecological Systematics* 5 (1974): 139–60; Doris Howes Calloway, ed., *Human Ecology in Space Flight* (New York: New York Academy of Sciences, 1966); idem, *Human Ecology in Space Flight II* (New York: New York Academy of Sciences, 1967); idem, *Human Ecology in Space Flight III* (New York: New York Academy of Sciences, 1968). See also Emanuel M. Roth, *Space-Cabin Atmospheres: A Literature Review*, 4 vols. (Washington, DC: NASA, 1964–67).

9. Stephen Bocking, *Ecologists and Environmental Politics: A History of Contemporary Ecology* (New Haven, CT: Yale University Press, 1997), 63–147; Joel B. Hagen, *An Entangled Bank: The Origins of Ecosystem Ecology* (New Brunswick, NJ: Rutgers University Press, 1992), 100–145; Donald Worster, *Nature's Economy: A History of Ecological Ideas*, 2nd ed. (Cambridge: Cambridge University Press, 1994), 365; Mark G. Madison, "'Potatoes Made of Oil': Eugene and Howard Odum and the Origins and Limits of American Agroecology," *Environment and History* 3 (1997): 209–38; Betty J. Craige, *Eugene Odum: Ecosystem Ecologist and Environmentalist* (Athens: University of Georgia Press, 2001), xi.

10. Frank B. Golley, "The Institute of Ecology," *Chemosphere* 4 (1975): 221–33; idem, "Establishing the Network," in *Holistic Science: The Evolution of the Georgia Institute of Ecology*, ed. Gary W. Barrett and Terry L. Barrett (New York: Taylor & Francis, 2001), 38–67; Bocking, *Ecologists and Environmental Politics*, 113.

11. Allan H. Brown summarizing Eugene Odum's views in Calloway, *Human Ecology in Space Flight*, 84.

12. Howard T. Odum, "Limits of Remote Ecosystems Containing Man," *American Biology Teacher* 25 (1963): 429–43, quotations on 430 and 442.

13. Mark Twain, *Life on the Mississippi* (Boston: James R. Osgood, 1883), 570; *Oxford English Dictionary*. See also Christian C. Young, "Defining the Range: Carrying Capacity in the History of Wildlife Biology and Ecology," *Journal of the History of Biology* 31 (1998): 61–83; and Sabine Höhler, "'Carrying Capacity'—the Moral Economy of the 'Coming Spaceship Earth,'" *Atenea* 26 (2006): 59–74.

14. Eugene Odum, quoted in Calloway, *Human Ecology in Space Flight*, 88.

15. Jack Myers, "Introductory Remarks," quotation on 409; Robert G. Tischer and Barbara P. Tischer, "Open Sequence Components of a Closed System," *American Biology Teacher* 25 (1963): 445–49.

16. Bernard C. Patten, "Information Processing Behavior of a Natural Plankton Community," *American Biology Teacher* 25 (1963): 489–501, quotation on 489. Cf. Geoffrey C. Bowker, "Biodiversity Datadiversity," *Social Studies of Science* 30 (2000): 643–83.

17. Eugene Odum, quoted in Calloway, *Human Ecology in Space Flight*, 87, 91; S. P. Johnson and J. C. Finn, "Ecological Considerations of a Permanent Lunar Base," *American Biology Teacher* 25 (1963): 529–35, figure on 530.

18. Eugene P. Odum, *Ecology* (New York: Holt, Rinehart & Winston, 1963), 10; Howard E. McCurdy, *The Space Station Decision: Incremental Politics and Technological Choice* (Baltimore: Johns Hopkins University Press, 1990).

19. Arthur C. Clarke, *2001: A Space Odyssey* (New York: Signet, 1968), 62, 44; Pierre Teilhard de Chardin, *The Future of Man*, trans. Norman Denny (London: Collins, 1964), 122.

20. Kim McQuaid, "Selling the Space Age: NASA and Earth's Environment, 1958–1990," *Environment and History* 12 (2006): 127–63.

21. Dennis Cooke, "Ecology of Space Travel," in *Fundamentals of Ecology*, ed. Eugene P. Odum, 3rd ed. (Philadelphia: W. B. Saunders, 1971), 498–509, quotations on 498, 509; Eugene Odum, quoted in Calloway, *Human Ecology in Space Flight*, 87; Ramón Margalef, *Perspectives in Ecological Theory* (Chicago: University of Chicago Press, 1968), 1.

22. Gerard K. O'Neill, "The Colonization of Space," *Physics Today* 27 (Sept. 1974): 32–40, quotations on 32; idem, "A Lagrangian Community?" *Nature* 250 (Aug. 1974): 636; idem, "Colonization of Space," interview by Richard M. Reis, *Mercury* 3 (July–Aug. 1974):

4-10. For a full review of the L-5 community, see De Witt Douglas Kilgore, *Astrofuturism* (Philadelphia: University of Pennsylvania Press, 2003), 150–85. On the Arcadian and managerial traditions in ecology, see Worster, *Nature's Economy*.

23. Gerard K. O'Neill, *The High Frontier* (New York: William Morrow, 1977), 201; idem, "Space Colonies: The High Frontier," *Futurist* 10 (1976): 25–33; Walter Sullivan, "Proposal for Human Colonies in Space Is Hailed by Scientists as Feasible," *New York Times*, 13 May 1974; Henry S. F. Cooper, *A House in Space* (New York: Macmillan, 1976); Michael A. G. Michaud, *Reaching for the High Frontier: The American Pro-Space Movement, 1972–84* (New York: Praeger, 1986); Gerard K. O'Neill, interview by James Metleager and Salvatore Napolitano, *Penthouse*, Aug. 1976, 87–90, 174–76 (I am grateful to Marilyn Peck Francescon for providing me with a copy of the interview).

24. Fred Turner, *From Counterculture to Cyberculture: Stewart Brand, the Whole Earth Network, and the Rise of Digital Utopianism* (Chicago: University of Chicago Press, 2006).

25. Brian O'Leary, "Mining the Apollo and Amor Asteroids," *Science*, 22 July 1977, 363–66; H. Keith Henson and Carolyn M. Henson, "Closed Ecosystems of High Agricultural Yield," in *Space Manufacturing Facilities*, ed. Jerry Grey (New York: American Institute of Aeronautics and Astronautics, 1977), 105–14; Carolyn M. Henson, "Space Agriculture Retort," *Co-Evolution Quarterly* 13 (1977): 50.

26. John C. Fletcher, foreword to *Space Settlements: A Design Study*, ed. Richard D. Johnson and Charles Holbrow (Washington, DC: National Aeronautics and Space Administration, 1977), v–vii, 181, fig. 5.15; Michael Modell, "Sustaining Life in a Space Colony," *Technological Review* 79 (Aug. 1977): 36–43. Some of the same ideas were expressed in Ernest Callenbach, *Ecotopia* (London: Pluto, 1978).

27. Erik Bergaust, *Colonizing Space* (New York: G. P. Putnam's Sons, 1978), 47. See also Edward Crowley, "Designing the Space Colony," *Technological Review* 79 (Aug. 1977): 45–50.

28. Gerard K. O'Neill, testimony in *Future Space Programs, 1975*, report prepared for the Subcommittee on Space Science and Applications of the House Committee on Science and Technology, 94th Cong., 1st sess., 111, modified version later published as Gerhard K. O'Neill, "Space Colonies and Energy Supply to the Earth," *Science*, 5 Dec. 1975, 943–47; "O'Neill testifies before Congress," *L-5 News*, Sept. 1975, 1; Donella H. Meadows et al., *The Limits to Growth* (New York: Universe Books, 1972); William Proxmire, quoted in *Spaceships of the Mind*, by Nigel Calder (New York: Viking, 1978), 40.

29. O'Neill, *High Frontier*, 32, 49, 97; Peter E. Glaser, "Power from the Sun," *Science*, 22 Nov. 1968, 857–61; Herman E. Daly, ed., *Toward a Steady-State Economy* (San Francisco: Freeman, 1973); idem, *Steady-State Economics* (San Francisco: Freeman, 1977).

30. O'Neill, *High Frontier*, 123. Building the space colony was reviewed as a "feasibility" by Donald Q. Innes in "Geographical Record," *Geographical Review* 68 (1978): 223–33, quotation on 223. An anonymous reviewer thought the book reviled "very real possibilities"; see Front Matter, *Social Forces* 57 (1978): 771–72 So did Ron Chernow, in "Colonies in Space May Turn Out to Be Nice Places to Live," *Smithsonian*, Feb. 1976, 62–69. For the theoretical basis of O'Neill's ideas, see Eugene P. Odum, "The Strategy of Ecosystem Development," *Science*, 18 Apr. 1969, 262–70; and Daniel S. Simberloff and Edward O. Wilson, "Experimental Zoogeography of Islands: The Colonization of Empty Islands," *Ecology* 50 (1969): 278–96.

31. Tom A. Heppenheimer, *Colonies in Space* (New York: Warner Books, 1977), 163, 186, emphasis in original. Heppenheimer's argument was based on J. Peter Vajk, "The Impact of Space Colonization on World Dynamics," *Technological Forecasting and Social Change* 9 (1976): 361–99.

32. Jerry Brown, "From Limits on Earth to Possibilities in Space," in *Space Colonies*, ed. Stewart Brand (San Francisco: Whole Earth Catalog, 1977), 146; Thomas R. Caudill, "Mass-Balance Model for a Controlled Ecological Life Support System on Mars," in *The Case for Mars II*, ed. Christopher P. McKay (San Diego: American Astronautical Society, 1985), 611–26; David T. Smernoff and Robert D. MacElroy, "Use of Martian Resources in a Controlled Ecological Life Support System," *Journal of the British Interplanetary Society* 42 (1989): 179–84; Yoji Ishikawa, Takaya Ohkita, and Yoji Amemiya, "Mars Habitation 2057," ibid. 43 (1990): 505–12.

33. Jack D. Salmon, "Politics of Scarcity versus Technological Optimism: A Possible Reconciliation?" *International Studies Quarterly* 21 (1977): 701–20; Hugh Albert Millward, "Geographical Aspects of the 'High Frontier' Concept," *Geografiska Annaler* 61 (1979): 113–21.

34. *Walt Disney World and EPCOT Center* (New York: Crescent Books, 1985).

35. See Gerard K. O'Neill, *2081: A Hopeful View of the Human Future* (New York: Touchstone, 1981); idem, *The Technology Edge* (New York: Simon & Schuster, 1983); Tom A. Heppenheimer, *The Real Future* (New York: Doubleday, 1983); Brian O'Leary, *The Fertile Stars* (New York: Everest House, 1981); idem, *Project Space Station* (Harrisburg, PA: Stackpole Books, 1983); Ben Bova, *The High Road* (Boston: Houghton Mifflin, 1981); idem, *Welcome to Moonbase* (New York: Ballantine Books, 1987); Jerry Grey, *Beachheads in Space* (New York: Macmillan, 1983); Marshall T. Savage, *The Millennial Project: Colonizing the Galaxy in Eight Easy Steps* (Boston: Little, Brown, 1994); Stanley Schmidt and Robert M. Zubrin, eds., *Islands in the Sky: Bold New Ideas for Colonizing Space* (New York: Wiley, 1996); Robert M. Zubrin, ed., *From Imagination to Reality: Mars Exploration Studies of the Journal of the British Interplanetary Society: Precursors and Early Piloted Exploration Missions* (San Diego: American Astronautical Society and the British Interplanetary Society by Univelt, 1997); and idem, *From Imagination to Reality: Mars Exploration Studies of the Journal of the British Interplanetary Society: Base Building, Colonization and Terraformation* (San Diego: American Astronautical Society and the British Interplanetary Society by Univelt, 1997). For favorable reviews of O'Neill's *2081*, see Alvin Rudoff, *Societies in Space* (New York: Peter Lang, 1996); and Robert Zimmerman, *Leaving Earth: Space Stations, Rival Superpowers, and the Quest for Interplanetary Travel* (Washington, DC: John Henry, 2003).

36. Frank B. Golley, "Environmental Ethics and Extraterrestrial Ecosystems," in *Beyond Spaceship Earth: Environmental Ethics and the Solar System*, ed. Eugene C. Hargrove (San Francisco: Sierra Club Books, 1986), 211–26, quotation on 224.

37. René Dubos, "Diagnosis of a General Sickness," undated interview by Anne Chisholm, in Chisholm, *Philosophers of the Earth* (New York: E. P. Dutton, 1972), 24.

38. Andrew Kirk, "Appropriating Technology: The Whole Earth Catalog and Counterculture Environmental Politics," *Environmental History* 6 (2001): 374–94, quotation on 389; for a review of the various editions of the *Whole Earth Catalog*, see 390n3. See also Andrew Kirk, *Counterculture Green: The "Whole Earth Catalog" and American Environmentalism* (Lawrence: University Press of Kansas, 2007); J. Baldwin and Stewart Brand,

eds., *Soft-tech* (New York: Penguin, 1978); and Turner, *From Counterculture to Cyberculture*.

39. Richard Buckminster Fuller, *Education Automation* (1962;(London: Jonathan Cape, 1973), 7.

40. John McHale, *The Future of the Future* (New York: George Braziller, 1969), 98, 106, 178.

41. Hugh Kenner, "We Will Get the Future We Learn to Expect," *New York Times*, 20 Apr. 1969; John Leonard, "Books of the Times," ibid., 21 Aug. 1969, 39.

42. William Bruneau and Mike Waters, "Game of the World," in *Kidstuff* (Berkeley, CA: Department of Architecture, University of California, 1969), 3–9 and cartoon on 48; Robert W. Driscoll, *Engineering Man for Space: The Cyborg Study*, NASA-512 (Farmingdale, CT: United Aircraft Corporate Center, 1963), reprinted in *Cyborg Handbook*, ed. Chris Hables Gray (London: Routledge, 1995), 76.

43. Donna Haraway, "A Manifesto for Cyborgs," *Socialist Review* 80 (1985): 65–108; idem, *The Companion Species Manifesto* (Chicago: Prickly Paradigm, 2003); Michel Serres, *The Natural Contract*, trans. Elizabeth MacArthur and William Paulson (Ann Arbor: University of Michigan Press, 1996), 121.

CHAPTER 7

1. James M. Fish, *American Building*, 2nd ed., vol. 2 (Boston: Houghton, 1972), viii.

2. Richard Buckminster Fuller, *Ideas and Integrities*, ed. Robert W. Marks (Englewood Cliffs, NJ: Prentice-Hall, 1963), 270. Richard Buckminster Fuller, *Operating Manual for Spaceship Earth* (Edwardsville: Southern Illinois University Press, 1969), 46. The first published mention of "spaceship earth" was in a children's book by Julius Schwartz, *The Earth Is Your Spaceship* (New York: McGraw Hill, 1963), a book probably inspired by Fuller.

3. Richard Buckminster Fuller, "City of the Future," *Playboy*, Jan. 1968, 166–68; Ulrich Franzen and Paul Rudolph, *The Evolving City* (New York: Whitney Library of Design for American Federation of Arts, 1974), 13.

4. Kenneth E. Boulding, "The Economics of the Coming Spaceship Earth," in *Environmental Quality in a Growing Economy*, ed. Henry Jarrett (Baltimore: Johns Hopkins Press, 1966), 3–14; idem, *Beyond Economics* (Ann Arbor: University of Michigan Press, 1968). The article was praised in Luther J. Carter, "Development in the Poor Nations: How to Avoid Fouling the Nest," *Science*, 7 Mar. 1969, 1046–48; and Darnell Rucker, "Beyond Economics" (review), *Ethics* 79 (1969): 243–44. It was criticized in J. W. Williams, "Environmental Quality" (review), *Journal of Business* 41 (1968): 268–69; and Luis Fernández-Galiano, *Fire and Memory: On Architecture and Energy*, trans. Gina Cariño (Cambridge, MA: MIT Press, 2000).

5. Barbara Ward, *Spaceship Earth* (New York: Columbia University Press, 1966), 14; Barbara Ward and René Dubos, *Only One Earth* (New York: Norton, 1972); Adlai Stevenson, "Strengthening the International Development Institutions," lecture before the U.N. Economic and Social Council, Geneva, July 1965, http://www.adlaitoday.org/article.php?id=6.

6. U Thant, "Statement at Dinner Inaugurating Twenty-Fifth Anniversary of United Nations Day Programme," *American Journal of International Law* 65 (1971): 447–52, quotation on 450.

7. Philippe de Seynes, "Prospect for a Future Whole World" (May 1971), reprinted in *International Organization* 26 (1972): 1–17, quotation on 1.

8. William T. R. Fox, "Science, Technology and International Politics," *International Studies Quarterly* 12 (1968): 1–15, quotation on 7. See also Luther J. Carter, "Earth Day: A Fresh Way of Perceiving the Environment," *Science*, 1 May 1970, 558–59; Isaac Asimov, *Earth: Our Crowded Spaceship* (London: Abelard-Schuman, 1974); Rodney F. Allen, Carmelo P. Foti, Daniel M. Ulrich, and Steven H. Woolard, *Deciding How to Live on Spaceship Earth* (Evanston, IL: McDougal, Littell, 1975); The Diagram Group, *Spaceship Earth: Its Voyage through Time* (New York: Hearst Books, 1980); and Nigel Calder, *Spaceship Earth* (London: Penguin, 1991).

9. John McHale, *The Ecological Context* (New York: George Braziller, 1970), 37, 166–74.

10. Johan McHale and Magda Cordell McHale, *Human Requirements, Supply Levels, and Outer Bounds* (New York: Aspen Institute for Humanistic Studies, 1975); idem, *The Futures Directory* (Boulder, CO: Westview, 1977).

11. Siegfred Fred Singer, "Spaceship Earth—A Global View of Ecology," in *Bioengineering and Cabin Ecology*, ed. W. B. Cassidy (Tarzana, AZ: American Association for the Advancement of Science, 1969), 1–7, quotation on 1; idem, ed., *Global Effects of Environmental Pollution* (New York: Springer Verlag, 1970); idem, *Is There an Optimum Level of Population?* (New York: McGraw-Hill, 1971).

12. Jack A. Kraft, "Industry's Utilization of Human Factors and Bioengineering," in Cassidy, *Bioengineering and Cabin Ecology*, 19–22, quotation on 22.

13. Siegfred Fred Singer, "The Research Potential of Manned Earth Orbiting Spacecraft for Meteorology," in *Manned Laboratories in Space*, ed. Singer (Dordrecht, Netherlands: Reidel, 1969), 33–45; W. T. Pecora, "Earth Resource Observation from an Orbiting Spacecraft," ibid., 75–87; Mark Kurlansky, *1968: The Year That Rocked the World* (New York: Ballantine Books, 2004), 381–83.

14. See Sheila Jasanoff, "Heaven and Earth: The Politics of Environmental Images," in *Earthly Politics: Local and Global in Environmental Governance*, ed. Jasanoff and Marybeth Long Martello (Cambridge, MA: MIT Press, 2004), 31–52, quotation on 32; Neil Maher, "Neil Maher on Shooting the Moon," *Environmental History* 9 (2004): 526–31; and Denis Cosgrove, "Contested Global Visions," *Annals of the Association of American Geographers* 84 (1994): 270–94.

15. Howard T. Odum, *Environment, Power, and Society* (New York: Wiley-Interscience, 1971), 125; idem, "Terminating Fallacies in National Policy on Energy, Economics, and Environment," in *Energy*, ed. Anton B. Schmalz (Washington, DC: World Future Society, 1975), 15–19; Howard T. Odum and Elisabeth C. Odum, *Energy Basis for Man and Nature* (New York: McGraw-Hill, 1976), 115–16; idem, *A Prosperous Way Down* (Boulder: University Press of Colorado, 2001); Eugene P. Odum, *Ecology and Our Endangered Life-Support Systems*, 2nd ed. (Sunderland, MA: Sinauer, 1992), 1–6.

16. James Lovelock to author, 24 Mar. 2007, author's personal archive.

17. James Lovelock, "A Physical Basis for Life Detection Experiments," *Nature* 207 (1965): 568–70; Dian R. Hitchcock and James E. Lovelock, "Life Detection by Atmospheric Analysis," *Icarus* 7 (1967): 149–59; Lynn Margulis and James Lovelock, "Biological Modulation of the Earth's Atmosphere," ibid. 21 (1974): 471–89; James Lovelock, *Gaia: A New Look at Life on Earth* (Oxford: Oxford University Press, 1979).

18. Paul R. Ehrlich and Richard L. Harriman, *How to Be a Survivor: A Plan to Save Spaceship Earth* (New York: Ballantine Books, 1971); Garrett Hardin, *Exploring New Ethics for Survival: The Voyage of the Spaceship Beagle* (New York: Viking, 1972); idem, "Living on a Lifeboat," *BioScience* 20 (Oct. 1974): 561–68; idem, "Carrying Capacity," in *Lifeboat Ethics*, ed. George R. Lucas and Thomas W. Ogletree (New York: Harper & Row, 1976), 135–36; James Lovelock, *Gaia: The Practical Science of Planetary Medicine* (London: Gaia Books, 1991), 153–55.

19. Dave Foreman and Bill Heywood, eds., *Ecodefence: A Field Guide to Monkeywrenching* (Tucson, AZ: Ned Ludd Books, 1987); Sale Kirkpatrick, "Deep Ecology and Its Critics," *Nation*, May 1988, 675; Christopher Manes, *Green Rage: Radical Environmentalism and the Unmaking of Civilization* (Boston: Little, Brown, 1990); Martha F. Lee, *Earth First! Environmental Apocalypse* (Syracuse, NY: Syracuse University Press, 1995).

20. Colin Moorcroft, ed., "Designing for Survival," special issue, *Architectural Design* 42 (July 1972): 413–45; Paul R. Ehrlich, *The Population Bomb* (New York: Ballantine Books, 1968); Donella H. Meadows et al., *The Limits to Growth* (New York: Universe Books, 1972); Lydia Kallipoliti, "Materials off the Catalogue," *Thresholds* 31 (2008): 9–17.

21. Kurlansky, *1968*, 382.

22. Dennis C. Pirages and Paul R. Ehrlich, *Ark II: Social Response to Environmental Imperatives* (New York: Viking, 1974), v. A particularly gloomy article is Paul Ehrlich, "Eco-Catastrophe!" in *Ecological Crisis: Readings for Survival*, ed. Glen A. Love and Rhoda M. Love (New York: Harcourt, 1970), 3–15.

23. Andrew Kirk, "Appropriating Technology: The *Whole Earth Catalog* and Counterculture Environmental Politics," *Environmental History* 6 (2001): 374–94; idem, "Machines of Loving Grace: Alternative Technology, Environment, and the Counterculture," in *Imagine Nation*, ed. Peter Braunstein and Michael W. Doyle (New York: Routledge, 2002), 353–78; Robert S. De Ropp, *Eco-Tech: The Whole-Earther's Guide to the Alternate Society* (New York: Delacorte, 1975).

24. Paolo Soleri, *The Omega Seed* (Garden City, NY: Anchor, 1981), 147–61.

25. Philip Steadman, *Energy, Environment and Building* (Cambridge: Cambridge University Press, 1975), 4. For a full review of alternative-energy research in this period, see Wilson Clark, *Energy for Survival: The Alternative to Extinction* (Garden City, NY: Anchor, 1974); Ralph L. Knowles, *Energy and Form: An Ecological Approach to Urban Growth* (Cambridge, MA: MIT Press, 1974); American Institute of Architects, *Energy*, 2 vols. (Washington, DC, 1975–80); and Ken Butti and John Perlin, *A Golden Thread: 2500 Years of Solar Architecture and Technology* (New York: Van Nostrand Reinhold, 1980).

26. [Colin Moorcroft], "Military," *Architectural Design* 42 (1972): 438.

27. Tim Ingold, "Globes and Spheres: The Topology of Environmentalism," in *Environmentalism: The View from Anthropology*, ed. Key Milton (London: Routledge, 1993), 31–42; Richard Grove, *Green Imperialism* (Cambridge: Cambridge University Press, 1995), 100–101.

28. Phillip Tabb, *Solar Energy Planning: A Guide to Residential Settlement* (New York: McGraw-Hill, 1984), xiii.

29. See the following by Constantinos A. Doxiadis: *Ecumenopolis* (Athens, Greece: Doxiadis Associates, 1963), 1; *The Two Headed Eagle: From the Past to the Future of Human Settlements* (Athens, Greece: Lycabuttus, 1972); *Building Entopia* (New York: Norton, 1975); *Ecology and Ekistics* (Boulder, CO: Westview, 1977).

30. Ian L. McHarg, *A Quest for Life: An Autobiography* (New York: Wiley, 1996), 79; idem, "The Court House Concept," *Architectural Record* 122 (Sept. 1957): 193–200; *The Classic McHarg: An Interview*, by E. Lynn Miller and Sidónio Pardal (Lisbon: CESUR, Technical University of Lisbon, 1992), 23–25; Christine Macy and Sarah Bonnemaison, *Architecture and Nature* (New York: Routledge, 2003), 137–221.

31. Ian L. McHarg, "Man and Environment," in *The Urban Condition*, ed. Leonard J. Duhl (New York: Basic Books, 1963), 44–58, quotation on 44. See also idem, "Regional Landscape Planning," in *Resources, the Metropolis, and the Land-Grant University* (New York: Conservation Foundation, 1963), 31–37.

32. Ian L. McHarg, "The Place of Nature in the City of Man," *Annals of the American Academy of Political and Social Science* 352 (Mar. 1964): 1–12, quotation on 1. See also Ian L. McHarg and David W. Wallace, "Plan for the Valleys vs. Spectre of Uncontrolled Growth," *Landscape Architecture* 55 (Apr. 1965): 179–81.

33. McHarg, *Quest for Life*, 132.

34. Jan Christian Smuts, *Holism and Evolution* (London: Macmillan, 1926); John Phillips, "The Biotic Community," *Journal of Ecology* 19 (1931): 1–24; Peder Anker, "The Politics of Ecology in South Africa on the Radical Left," *Journal of the History of Biology* 37 (2004): 303–31.

35. John Phillips, "Problems in the Use of Chemical Fertilizers," in *The Careless Technology: Ecology and International Development*, ed. M. Taghi Farvar and John P. Milton (Garden City, NY: Natural History Press, 1972), 549–66, quotation on 555.

36. John Phillips, "Ecology and the Ecological Approach," *Via* 1 (1968): 17–18.

37. Jack McCormick, "Succession," ibid., 22–35; Louis Kahn, "Silence," ibid., 88–89; Aldo van Eyck, "Design Only Grace," ibid., 102–15; Fritz Morgenthaler, "The Dogon People," ibid., 116–23; Nicholas Muhlenberg, "Ecology, Economics, and Planning," ibid., 19–21. See also Ruth Patrick, "Natural and Abnormal Communities of Aquatic Life in Streams," ibid., 36–41.

38. Ian L. McHarg, "Ecology, for the Evolution of Planning and Design," ibid., 44–67, quotation on 66.

39. Ian L. McHarg, *Design with Nature* (Garden City, NY: Doubleday, 1969), iv; idem, *Quest for Life*, 331–32.

40. See McHarg, *Design with Nature*, 2, 28–29; and Edward W. Said, *Orientalism* (New York: Pantheon Books, 1978).

41. McHarg, *Design with Nature*, 95.

42. Ibid., 46, 96, 97.

43. Ibid., 98.

44. Ibid., 101, 197.

45. Ian L. McHarg, "Values, Process and Form," in *The Fitness of Man's Environment*, with an introduction by Jennie Lee (Washington, DC: Smithsonian Institution Press, 1968), 207–27, quotation on 209; idem, "An Ecological Method for Landscape Architecture," *Landscape Architecture* 57 (Feb. 1967): 105–7; idem, "Where Should Highways Go?" ibid. 57 (Apr. 1967): 179–81; idem, "Ecological Determinism," in *Future Environments of North America*, ed. F. Fraser Darling and John P. Milton (Garden City, NY: Natural History Press, 1966), 526–38.

46. McHarg, *Quest for Life*, 203, 206. For a favorable review, see, for example, Diane L. Ringger and Forest Stearns, "Nature's Landscape Architect," *Ecology* 51 (Nov. 1970): 1109–10.

47. Ian L. McHarg, "Architecture in an Ecological View of the World," *AIA Journal* 54 (Nov. 1970): 47–51, quotation on 48.

48. Ibid., 49, 50. See also Ian L. McHarg, "Open Space from Natural Processes," in *Metropolitan Open Space and Natural Process*, ed. David A. Wallace (Philadelphia: University of Pennsylvania Press, 1970), 10–52.

49. Ian L. McHarg, "The Environmental Crisis," in "The Consequences of Today," special issue, *Architecture in Australia* 59 (Aug. 1971): 638–46, quotation p. 638.

50. Ian L. McHarg and Jonathan Sutton, "Ecological Plumbing for the Texas Coastal Plain," *Landscape Architecture* 65 (Jan. 1975): 78–89; Ian L. McHarg, "Must We Sacrifice the West?" in *Growth Alternatives for the Rocky Mountain West*, ed. Terrell J. Minger and Sherry D. Oaks (Boulder, CO: Westview Press, 1976), 203–11; idem, "Biological Alternatives to Water Pollution," in *Biological Control of Water Pollution*, ed. Joachim Tourbier and Robert W. Person (Philadelphia: University of Pennsylvania Press, 1976), 7–12; Arthur H. Johnson, Jonathan Berger, and Ian L. McHarg, "A Case Study in Ecological Planning: The Woodlands, Texas," in *Planning the Uses of Management of Land*, ed. Marvin T. Beatty, Gary W. Petersen, and Lester D. Swindale (Madison, WI: American Society of Agronomy, 1978), 935–56.

51. John McHale and Magda Cordell McHale, *Basic Human Needs: A Framework for Action* (New Brunswick, NJ: Transaction Books, 1978).

52. John Todd, "Comments," in *Space Colonies*, ed. Stewart Brand (San Francisco: Whole Earth Catalog, 1977), 48–49; Tom A. Heppenheimer, *Colonies in Space* (New York: Warner Books, 1977), 170, 180.

53. Antonio Ballester, Daniel B. Botkin, James Lovelock, Ramón Margalef, Lynn Margulis, Juan Oro, Rusty Schweikert, David Smith, T. Swain, John Todd, Nancy Todd, and George M. Woodwell, "Ecological Considerations for Space Colonies," *Bulletin of the Ecological Society of America* 57 (1976): 2–4, quotation on 3; also published in *Co-Evolution Quarterly* 12 (1976): 96–97, reprinted in Brand, *Space Colonies*, 92–93.

54. John Todd and Nancy Todd, *Tomorrow Is Our Permanent Address: The Search for an Ecological Science of Design as Embodied in the Bioshelter* (New York: Harper & Row, 1980), 33; Nigel Calder, *Spaceships of the Mind* (New York: Viking, 1978), 61–63; Pierre Elliot Trudeau, "From Urgencies to Essentials," *Co-Evolution Quarterly* 12 (1976–77): 102–3; J. Baldwin, "The New Alchemists Are neither Magicians nor Geniuses. They Are Hard Workers," ibid., 104–11.

55. John L. Hess, "Farm-Grown Fish," *New York Times*, 6 Sept. 1973; John Todd, "Pioneering for the 21st Century: A New Alchemist's Perspective," *Ecologist* 6 (1976): 252–57.

56. Betty Roszak, foreword to *The Book of the New Alchemist*, ed. Nancy Todd and John Todd (New York: E. P. Dutton, 1977), vii.

57. William O. McLarney and John Todd, "Walton Tow: A Complete Guide to Backyard Fish Farming," in ibid., 74–106, quotation on 74. See also John Todd, "The Ark: A Solar-Heated, Wind-Powered Greenhouse and Fish Pond Complex," *Futurist* 8 (1974): 296–98.

58. Nicholas Wade, "New Alchemy Institute: Search for an Alternative Agriculture," *Science*, 28 Feb. 1975, 727–29, quotation on 727; idem, "Limits to Growth: Texas Conference Finds None, but Didn't Look Too Hard," ibid., 7 Nov. 1975, 540–41.

59. Wade Green, "The New Alchemy," *New York Times*, 8 Aug. 1976.

60. John Todd, "A Modest Proposal: Science for the People," in *Radical Agriculture*, ed.

Richard Merrill (New York: New York University Press, 1976), 259–83, quotation on 270; Allan L. Hammon, "Individual Self-Sufficiency in Energy," *Science*, 19 Apr. 1974, 278–82.

61. Nicholas Wade, "Windmills: The Resurrection of an Ancient Energy Technology," *Science*, 7 June 1974, 1055–58.

62. New Alchemy Institute West, "Methane Gas Digesters for Fuel and Fertilizer," in *Producing Your Own Power*, ed. Carol Hupping Stoner (Emmaus, PA: Rodale, 1974), 137–76; Daniel Shreeve, "Producing Your Own Power," *Quarterly Review of Biology* 51 (1976): 355–56.

63. James K. Page Jr. and Wilson Clark, "The New Alchemy: How to Survive in Your Space Time," *Smithsonian*, Feb. 1975, 82–89, quotation on 84.

64. In 1979 the New Alchemists charged an admission fee of two dollars. In comparison, the nearby Falmouth Historical Society Museum charged only one dollar. By 1982 the fee was three dollars. Phyllis Meras, "What's Doing on Upper Cape Cod," *New York Times*, 22 July 1979; Susan Daar, "'Ark' on the Cape," ibid., 15 Aug. 1982.

65. Green, "New Alchemy."

66. William O. McLarney, "Aquaculture: Toward an Ecological Approach," in Merrill, *Radical Agriculture*, 328–39.

67. Dean E. Abrahamson, "The Energy Crisis," *Science*, 19 Nov. 1971, 857–58; Hammon, "Individual Self-Sufficiency in Energy."

68. Nancy Todd and John Todd, *From Eco-Cities to Living Machines: Principles of Ecological Design*, rev. ed. (Berkeley, CA: North Atlantic Books, 1994), 167–69; idem, "Lessons from the Biosphere," *Design Aid*, Sept.–Oct. 1987, 56–59.

69. John Todd, "Ocean Arks," *Co-Evolution Quarterly* 23 (1979): 46–55, quotation on 46; J. Baldwin, "Trials of an Ocean Ark Model," ibid. 24 (1979–80): 56–57; John Todd, "Ocean Ark Corp.," ibid., 57–59; Nancy Todd and John Todd, *Bioshelters, Ocean Arks, City Farming: Ecology as the Basis of Design* (San Francisco: Sierra Club Books, 1984), 34–35; Carl H. Hertel, review of *Bioshelters, Ocean Arks, City Farming*, by Nancy Todd and John Todd, *Quarterly Review of Biology* 60 (1985): 556–57.

70. Michael Reynolds, *Earthship*, 3 vols. (Taos, NM: Solar Survival Press, 1990), 1:v–vi, emphasis in original.

CHAPTER 8

1. Stewart Brand, "Comments on O'Neill's Space Colonies," in *Space Colonies*, ed. Brand (San Francisco: Whole Earth Catalog, 1977), 33; Gerard O'Neill, "The High Frontier," "Testimony," and interview by Brand, reprints from, *Co-Evolution Quarterly* 2 (Fall 1975): 8–11, 12–21, 22–30.

2. Richard Buckminster Fuller, "Comment," in Brand, *Space Colonies*, 55; idem, "City of the Future," *Playboy*, Jan. 1968, 166–68; Wolf von Eckardt, "That Bucky Fuller Sure Is a High Flyer," *Washington Post*, 2 Oct. 1966.

3. Carl Sagan, "Comment," in Brand, *Space Colonies*, 42.

4. Tom A. Heppenheimer, "Space Agriculture and Space Cops," in ibid., 94–95.

5. Jesco von Puttkamer, "Driving or Driven to Space?" *Co-Evolution Quarterly* 11 (1976): 46.

6. H. Keith Henson, "8am of the World," ibid., 50.

7. "Jacques Cousteau at NASA Headquarters," interview by Stewart Brand, in Brand, *Space Colonies*, 98–103. For a full discussion of the importance of satellite technologies

for monitoring Earth, see Frank White, *The Overview Effect: Space Exploration and Human Evolution* (Boston: Houghton Mifflin, 1987).

8. Lynn Margulis, "Comments," in Brand, *Space Colonies*, 35; idem, ed., *Origins of Life* (New York: Springer Verlag, 1973).

9. Lewis Mumford to Stewart Brand, n.d., reprinted in Brand, *Space Colonies*, 34; Lewis Mumford, *The Myth of the Machine: The Pentagon of Power* (New York: Harcourt, 1970).

10. Ken Kesey, "Comment," in Brand, *Space Colonies*, 34.

11. Gary Snyder, "Comment," in ibid., 69.

12. Ernst F. Schumacher, "Comment," in ibid., 38.

13. Wilson Clark, "Comment"; Dennis Meadows, "Comment"; Garrett Hardin, "Comment," in ibid., 38, 40, and 54.

14. Paul R. Ehrlich and Anne H. Ehrlich, "Comment," in ibid., 43; Paul R. Ehrlich, Anne H. Ehrlich, and John P. Holdren, *Ecoscience: Population, Resources, Environment* (San Francisco: Freeman, 1977), 821–22.

15. George Wald, "Comment," in Brand, *Space Colonies*, 44–45.

16. Wendell Berry, "Comment," in ibid., 36–37, quotation on 36; idem, "The Debate Sharpens," ibid., 82–84.

17. David Bowie, "Space Oddity," *Space Oddity*, Phillips SBL 7912.

18. Paolo Soleri, "Comment," in Brand, *Space Colonies*, 56–60. Soleri later became a supporter of space colonization; see his *The Omega Seed* (Garden City, NY: Anchor, 1981), 147–61.

19. William Irwin Thompson, "Comment," in Brand, *Space Colonies*, 44.

20. John Todd, "Comments," in ibid., 48–49; Tom A. Heppenheimer, *Colonies in Space* (New York: Warner Books, 1977), 170, 180.

21. Antonio Ballester, Daniel B. Botkin, James Lovelock, Ramón Margalef, Lynn Margulis, Juan Oro, Rusty Schweikert, David Smith, T. Swain, John Todd, Nancy Todd, and George M. Woodwell, "Ecological Considerations for Space Colonies," *Bulletin of the Ecological Society of America* 57 (1976): 2.

22. "Grumman's Integrated Household System," *Architectural Design* 42 (July 1972): 423; Richard Thruelsen, *The Grumman Story* (New York: Praeger, 1976), 292, 325.

23. Jack A. Kraft, "Industry's Utilization of Human Factors and Bioengineering," in *Bioengineering and Cabin Ecology*, ed. W. B. Cassidy (Tarzana, AZ: American Association for the Advancement of Science, 1969), 19–22, quotation on 22.

24. Helga Olkowski et al., *The Integral Urban House* (San Francisco: Sierra Club Books, 1979), xviii. See also Berkeley Tribe, "Blueprint for a Communal Environment," in *Sources*, ed. Theodore Roszak (New York: Harper & Row, 1972), 392–413.

25. Sean Wellesley-Miller and Day Chahroudi, "Bio Shelter," *Architecture Plus* 2 (1974): 90–95, quotations on 92.

26. Sean Wellesley-Miller and Day Chahroudi, "Buildings as Organisms," *Architectural Design* 45 (1975): 157–62, quotation on 157; Nancy Todd, "Bioshelters and Their Implications for Lifestyle," *Habitat International* 2 (1977): 87–100.

27. Alexander Pike, "Cambridge Studies," *Architectural Design* 42 (July 1972): 441–45, quotation on 441, image on 442. Brenda Vale and Robert Vale, *The New Autonomous House* (London: Thames & Hudson, 2000), 8.

28. Janine Clarke and Robin Clarke, "The Philosophy and Aims of the Proposed Community," *Undercurrents*, Jan. 1972, reprinted in Robert Vale, *Services for an Autonomous*

Research Community in Wales (Cambridge: Cambridge University, Department of Architecture, 1974), 3–7; Robert Vale, *Analysis of Forms for an Autonomous House* (Cambridge: Cambridge University, Department of Architecture, 1973).

29. Brenda Vale and Robert Vale, *The Autonomous House: Design and Planning for Self-Sufficiency* (London: Thames & Hudson, 1975), 8, 184.

30. Ibid., 18.

31. Brenda Vale and Robert Vale, *Towards a Green Architecture* (London: RIBA, 1991); idem, *Green Architecture* (London: Thames & Hudson, 1991); idem, *New Autonomous House*.

32. Michelle Murphy, *Sick Building Syndrome and the Problem of Uncertainty: Environmental Politics, Technoscience, and Women Workers* (Durham, NC: Duke University Press, 2006), 131–50.

33. Kenneth Yeang, *A Theoretical Framework for Incorporating Ecological Considerations in the Design and Planning of the Built Environment* (PhD diss., Cambridge University, Department of Architecture, 1980). The thesis was edited and republished as *Designing with Nature: The Ecological Basis for Architectural Design* (New York: McGraw-Hill, 1995). See also by Yeang: *Rethinking the Environmental Filter* (Singapore: Landmark Books, 1989); and *The Architecture of Malaysia* (Amsterdam: Pepin, 1992), 19.

34. Kenneth Yeang, "Bionics: The Use of Biological Analogies for Design," *Architectural Association Quarterly* 6 (1974): 48–57, quotation on 48. Yeang's chief source of inspiration was Henrich Hertel, *Structure—Form—Movement* (New York: Reinhold, 1963). See also John Frazer, *An Evolutionary Architecture* (London: Architectural Association, 1995).

35. Kenneth Yeang, "Bases for Ecosystem Design," *Architectural Design* 42 (1972): 434–36, quotations on 435 and 436.

36. Kenneth Yeang, "Energetics of the Built Environment," ibid. 44 (1974): 446–51.

37. Kenneth Yeang, *Tropical Urban Regionalism* (Singapore: Concept Media, 1987), 41; idem, *The Tropical Verandah City* (Ehsan, Malaysia: Longman, 1987); idem, *Malaysia: Bioclimatic Skyscrapers* (Berlin: Aedes, 1994); idem, *Bioclimatic Skyscrapers* (London: Artemis, 1994).

38. Yeang, *Designing with Nature*, 61. Yeang's chief inspirations were Howard T. Odum, "Limits of Remote Ecosystems Containing Man," *American Biology Teacher* 25 (1963): 429–43; and idem, *Environment, Power, and Society* (New York: Wiley-Interscience, 1971).

39. Yeang, *Designing with Nature*, 61.

40. Kenneth Yeang, *The Skyscraper Bioclimatically Considered* (London: Academy, 1996); idem, *The Green Skyscraper* (Munich: Prestel, 1999); idem, "Bruno Stagno," in *Bruno Stagno: An Architect in the Tropics*, ed. Alexander Tzonis, Liane Lefaivre, and Kenneth Yeang (Ehsan, Malaysia: Asia Design Forum, 1999), 30–31; idem, *Service Cores* (Chichester, UK: Wiley, 2000); Leon van Schaik, *Ecocells* (Chichester, UK: Wiley, 2003); Kenneth Yeang, *Reinventing the Skyscraper: Vertical Theory of Urban Design* (Chichester, UK: Wiley, 2002); T. R. Hamzah and Kenneth Yeang, *T. R. Hamzah & Yeang: Selected Works* (Victoria, Australia: Images Publication Group, 1998).

41. [Colin Moorcroft], "Experiments with Power/Food/Water Systems," *Architectural Design* 42 (1972:, 424–25.

42. See Mark Nelson, Matt Finn, Cherie Wilson, Bernd Zabel, Mark van Thillo, Philip Hawes, and Rodrigo Fernandez, "Bioregenerative Recycling of Wastewater in Biosphere 2 Using a Constructed Wetland: 2-Year Results," *Ecological Engineering* 13 (1999): 189–97.

43. Kevin Kelly, *Out of Control: The Rise of Neo-Biological Civilization* (New York: William Patrick, 1994), 138; Timothy Miller, "The Sixties-Era Communes," in *Imagine Nation*, ed. Peter Braunstein and Michael W. Doyle (New York: Routledge, 2002), 327–51; Marina Benjamin, *Rocket Dreams* (New York: Free Press, 2003), 138–39; John Allen, *Biosphere 2: The Human Experiment* (New York: Penguin, 1991), 2–3.

44. "Bass on Biosphere 2 and 'Walden Two,'" *New York Times*, 24 Sept. 1991; Mark Nelson and Gerald Soffen, eds., *Biological Life Support Technologies: Commercial Opportunities*, NASA Conference Publication 3094 (Washington, , DC: National Aeronautics and Space Administration, 1990); Stewart Brand, *The Media Lab: Inventing the Future at MIT* (New York: Viking, 1987); J. Baldwin and Stewart Brand, eds., *Soft-tech* (New York: Penguin, 1978).

45. Thomas O. Paine, "Biospheres and Solar System Exploration," in Nelson and Soffen, *Biological Life Support Technologies*, 1–11, quotation on 1.

46. Dorion Sagan and Lynn Margulis, *Biospheres: From Earth to Space* (Hillside, NJ: Enslow, 1989), 11, 12, 38, 42–43, 61, 70, 85. See also Lynn Margulis and Oona West, "Gaia and the Colonization of Mars" (1993), in *Slanted Truths*, ed. Lynn Margulis and Dorion Sagan (New York: Springer Verlag, 1997), 221–34; Nigel Calder, *Spaceships of the Mind* (New York: Viking, 1978), 43–53; John Allen and Mark Nelson, *Space Biospheres* (Oracle, AZ: Synergetic Press, 1986); and idem, "Biospherics and Biosphere 2, Mission One (1991–1993)," *Ecological Engineering* 13 (1999): 15–29.

47. Robert J. Beyers and Howard T. Odum, *Ecological Microcosms* (New York: Springer Verlag, 1993), 397; John P. Allen, "Historical Overview of the Biosphere 2 Project," in Nelson and Soffen, *Biological Life Support Technologies*, 12–22; Robert E. Kohler, *Landscapes and Labscapes* (Chicago: University of Chicago Press, 2002).

48. Linnea Gentry and Karen Liptak, *The Glass Ark: The Story of Biosphere 2* (New York: Puffin Books, 1991), 7; Lee Durrell, *Gaia: State of the Ark Atlas* (New York: Doubleday, 1986), 184–211.

49. Carlyle C. Douglas, "A Voyage of Discovery That Doesn't Move," *New York Times*, 29 Sept. 1991; William J. Broad, "Recycling Claim by Biosphere 2 Experiment Is Questioned," ibid., 12 Nov. 1991; "Air Is Pumped into Biosphere 2," ibid., 20 Dec. 1991; "Outside Air Added to Biosphere Experiment," ibid., 21 Dec. 1991; Abigail Alling and Mark Nelson, *Life under Glass: The Inside Story of Biosphere 2* (Oracle, AZ: Biosphere Press, 1993).

50. Traci Watson, "Can Basic Research Ever Find a Good Home in Biosphere 2?" *Science*, 19 Mar. 1993, 1688–89; Tim Appenzeller, "Biosphere 2 Makes a New Bid for Scientific Credibility," ibid., 11 Mar. 1994, 1368–69; Kelly, *Out of Control*, 139.

51. Seth Mydans, "8 Bid Farewell to the 'Future,'" *New York Times*, 27 Sept. 1993; "Biospherics," editorial, ibid., 29 Sept. 1993; William J. Broad, "Too Rich a Soil," ibid., 5 Oct. 1993.

52. "Financial Backer Ousts Biosphere's Top Officers," ibid., 2 Apr. 1994; "A New Crew and New Rules for Biosphere 2," ibid., 20 Feb. 1994; "Two Former Biosphere Workers Are Accused of Sabotaging Dome," ibid., 5 Apr. 1994; Drummond Ayres, "Ecological Experiment Becomes Battleground," ibid., 11 Apr. 1994.

53. Beyers and Howard T. Odum, *Ecological Microcosms*, 3, 10.

54. John B. Corliss, "Ecosystems Lessons," *New York Times*, 12 Oct. 1993. For supportive statements, see www.biospheres.com/peoplepressquotes.html; Mark Nelson

and William F. Dempster, "Living in Space: Results from Biosphere 2's Initial Closure, an Early Testbed for Closed Ecological Systems on Mars," in *Strategies for Mars*, ed. Carol R. Stoker and Carter Emmart (San Diego: American Astronautical Society, 1996), 363–90; and Wallace S. Broecker, "The Biosphere and Me," *GSA Today* 6 (1996): 2–7.

55. Christopher P. McKay, "Terraforming Mars," *Journal of the British Interplanetary Society* 35 (1982): 427–33; David J. Thomas, "Biological Aspects of the Ecopoeisis and Terraformation of Mars," ibid. 48 (1995): 415–18; idem, "The Formation of Martian Ecosystems," in *The Case of Mars VI*, ed. Kelly R. McMillen (San Diego: American Astronautical Society, 2000), 445–51; Julian A. Hiscox, "Biology and the Planetary Engineering of Mars," in ibid., 453–81.

56. John Allen, "The Cosmic Drama," in *Design Outlaws on the Ecological Frontier*, ed. Chris Zelov and Phil Cousineau (Easton, PA: Knossus, 1997), 162–65; John Todd, "The New Alchemists," in ibid., 172–83.

57. Eugene Odum, "Biosphere 2: New Kind of Science," *Science*, 14 May 1993, 878–79.

58. Beyers and Howard T. Odum, *Ecological Microcosms*, 419, 427, 431; Abigail Alling, Mark Nelson, Linda Leigh, Taber MacCallum, Norberto Alvarez-Romo, and John Allen, "Experiments in Closed Ecological System in the Biosphere 2 Test Module," in ibid., 463–79; Howard T. Odum and Elisabeth C. Odum, *Modeling for All Scales* (San Diego: Academic Press, 2000).

59. Bruno D. V. Marion and Howard T. Odum, "Biosphere 2: Introduction and Research Progress," *Ecological Engineering* 13 (1999): 3–14, quotation on 12.

CONCLUSION

1. Frederic Towndrow, *Architecture in the Balance: An Approach to the Art of Scientific Humanism* (New York: Frederick Stokes, 1934), 139. See also Reyner Banham's classic *Architecture of the Well-Tempered Environment* (Chicago: University of Chicago Press, 1969).

2. Sam Love, "The Overconnected Society," *Futurist* 8 (1976): 293–95, quotation on 294.

3. Chris Zelov and Phil Cousineau, eds., *Design Outlaws on the Ecological Frontier* (Easton, PA: Knossus, 1997).

4. William McDonough and Michael Braungart, *Cradle to Cradle: Remaking the Way We Make Things* (New York: North Point, 2002), 65.

5. Richard Neutra, *World and Dwelling* (New York: Universe Books, 1962), 26.

6. Moshe Safdie, *Beyond Habitat* (Cambridge, MA: MIT Press, 1970); Judith Wolin, ed., *For Everyone a Garden* (Cambridge, MA: MIT Press, 1974).

7. Malcolm B. Wells, "An Ecologically Sound Architecture Is Possible," *Architectural Design* 42 (1972): 433–34; idem, *Underground Designs* (Brewster, : published by the author, 1977); idem, *Gentile Architecture* (New York: McGraw-Hill, 1981).

8. Hassan Fathy, *Natural Energy and Vernacular Architecture*, ed. Walter Shearer and Abd-el-rahman Ahmed Sulatan (Chicago: University of Chicago Press, 1986); Jean Detier, *Down to Earth: Mud Architecture* (London: Thames & Hudson, 1982); Athena Swentzell Steen, Bill Steen, and David Bainbridge, *The Straw Bale House* (White River Junction, VT: Chelsea Green, 1994); John Connell, *Home Instinct* (New York: McGraw-Hill, 1998); Pat Borer and Cindy Harris, *The Whole House Book: Ecological Building Design and Materi-*

als (Machynlleth, UK: Centre for Alternative Technology, 1998); Sydney Baggs and Joan Baggs, *The Healthy House* (Sydney, Australia: HarperCollins, 1996); David Pearson, *The Natural House Book: Creating a Healthy, Harmonious, and Ecologically-Sound Home Environment* (New York: Fireside, 1989).

9. Brian Edwards, ed., *Green Buildings Pay* (London: Spon, 1998); George W. Reinberg, *Architecture by George W. Reinberg* (Florence: Alinea, 1998); Union Internationale des Architects, *Eco Logical Architecture* (Stockholm, 1992); Tom Woolly, Sam Kimmins, Paul Harrison, and Rob Harrison, *Green Building Handbook*, 2 vols. (London: Spon, 1997–2000); Donald Watson, ed., *The Energy Design Handbook* (Washington, DC: American Institute of Architects Press, 1993); Sim Van der Ryn and Stuart Cowan, *Ecological Design* (Washington, DC: Island Press, 1996); Charles Jencks, *The Architecture of the Jumping Universe* (West Sussex, UK: Academy, 1997); John Hermannson, *Green Building Resource Guide* (Newton, CT: Taunton, 1997); *Energy Efficient Houses* (Newtown, CT: Taunton, 1993); National Audubon Society, *Audubon House: Building the Environmentally Responsible, Energy-Efficient Office* (New York: Wiley, 1994).

10. Clark Snell and Tim Callahan, *Building Green* (New York: Lark Books, 2005).

11. Shigeru Ban, *Shigeru Ban* (Hannover, Germany: Expo 2000, 1999); Winy Maas, *Metacity-Datatown* (Rotterdam: MVRDV/010, 1999).

12. Randall Stout Architects, *Environmental Alchemy* (New York: Edizioni, 2004).

13. Michael Hensel, Achim Menges, and Michael Weinstock (from the Ocean Group), eds., "Emergence: Morphogenetic Design Strategies," special issue, *Architectural Design* 74 (May–June 2004); Michael Hensel and Achim Menges, eds., *Morpho-Ecologies* (London: Architectural Association, 2006); idem, eds., "Versatility and Vicissitude: Performance in Morpho-Ecological Design," special issue, *Architectural Design* 78 (Mar.–Apr. 2008); Hughe Aldersey-Williams, "Towards Biomimetic Architecture," *Nature Materials* 3 (2004): 277–79; Foreign Office Architects, *Phylogenesis: FOA's Ark* (Barcelona: Actar, 2004); Dennis Dollens, *Digital-Botanic Architecture* (Santa Fe, NM: SITES Books, 2004).

14. Julian Huxley, quoted in *Gropius*, by Reginald R. Isaacs (Boston: Little, Brown, 1991), 295.

INDEX

Abbey, Edward, 8
Abercrombie, Patrick, 12
Abu Dhabi, 130
Adams, Ansel, 57
Advertisements, 55-56, 62, *following p. 82*
After Democracy (Wells), 148n41
Agriculture, 91, 92-93, 107, 109-10, 114, 121
AIDS, 101
Air Force, U.S., 71, 117
Aircraft (Le Corbusier), 32
Airplanes, 51-53, 70
"Airways to Peace" exhibition, 51-53, 60
Albers, Josef, 4
Alexander, Christopher, 39-40, 133
All Aboard for Ararat (Wells), 36
Allen, James S., 41
Allen, John, 122
American Airlines, 52
American Astronautical Society, 98-99
American Institute of Architects, 108
American Institute of Biological Sciences, 84, 85
Anderson, Garry, 67
Animal Biology (Haldane and J. Huxley), 23
Apollo program, 99, 117, 119. *See also* Space exploration and colonization
Arab oil embargo (1973-74), 102-3
Arcadian tradition in ecology, 4, 6, 8, 90

Architecture: McHarg on, 108; Moholy-Nagy's definition of, 16; and public health, 12, 21, 28; relationship between science and generally, 7. *See also* Bauhaus school; Ecological design
Architecture and the Sciences (Picon and Ponte), 7
Architecture Here and Now (Williams-Ellis and Summerson), 12
Architecture of Science (Galison and Thompson), 7
Arizona Biosphere 2, 113, 116, 121-25, 127
Ark II (Ehrlich, Harriman, and Pirages), 102
Ark projects, 102, 109-12, 121-25. *See also* Noah's Ark
Arup, Johannes, 21, 133
Arup, Ove, 10, 19, 21
Aspen, Colo., 55-59, 66, 126
Astronauts. *See* Space exploration and colonization
Atlases, 54-55, 58-66, *following p. 82*
Atomic Energy Commission, 85
Atomic weapons and atomic warfare, 71, 73-74, 85, 89. *See also* Cold war
Augustine, Margret, 122
Autonomous design. *See* Closed ecosystems

177

Balfour, Isaac, 104
Ban, Shigeru, 130
Banham, Reyner, 13
Bank of America building (New York), 130
Barr, Alfred, 18
Barry, Gerald, 25
Bass, Edward P., 122, 124
Bauhaus school: and Aspen, Colo., development, 55-58, 126; and biological sciences, 2-5, 13-23, 126-27; and film *Things to Come*, 29-36, 41; in Germany, 48-50; and global perspective, 48-53; goals of, 41, 51, 58, 67, 126-27, 128, 131; and housing standards, 25-26; and humanism generally, 4; and Isokon Building, 10-11; in London, 4, 9-23, 30, 126; and London Zoo, 18-23, 26, 29; and MARS (Modern Architectural Research Group), 10-12, 20; meaning of term, 4; Museum of Modern Art exhibition on, 37, 50-51, 52; in United States, 36-53, 126; and Wells, 12-13. *See also* Bayer, Herbert; Gropius, Walter; Moholy-Nagy, László; and other architects
Bayer, Herbert: and "Airways to Peace" exhibition, 51-53, 60; in Aspen, Colo., 55-59, 126; and Bauhaus school, 4, 9, 37, 48-67; biographical information on, 49, 133; and cartography, 54-55, 58-67, *following p. 82*; and Container Corporation advertisements, 56, *following p. 82*; designs with nature by, 54, 55-58; on energy, 64; environmental symbols created by, 61; and environmentalism, 60, 64-67; and functionalism, 50, 51; on geography, 58-59; global design of, 48-53; graphic design of, 54-67; *Grass Mound* by, 54, 58; health problems of, 57; and Museum of Modern Art exhibitions, 50-53, 60; paintings by, 17; patron of, 55-57, 59; on population growth, 60, 61; sgraffito mural by, 57-58; typography created by, 50; in United States, 37, 50-67; *World Geo-Graphic Atlas* by, 54-55, 58-66, *following p. 82*

Bayer, Joella, 57
Bernal, John D., 7, 19, 89-90
Berry, Wendell, 115-16
Beyers, Robert, 123, 125
Bildungsideal, 41
Bioclimatic design, 11, 120
Biological sciences: and Bauhaus school, 2-5, 13-23, 126-27; and bio-technique/bionics, 15-16, 29, 31, 32, 42, 50, 120, 131; and Biosphere 2 in Arizona, 122; evolutionary biology, 21-23, 27, 108, 120; and London Zoo, 18-23, 26, 29; and Moholy-Nagy, 13-18, 41-42, 50, 70, 131; Moholy-Nagy's biological bill of rights, 41; space program and biotechnologies, 86-87, 93-95, 127, 128. *See also* Ecological design
Biomimetics, 130
Bio-shelter, 118
Biosphere of Earth, *following p. 82*, 99, 121
Biosphere 2 (Arizona), 113, 116, 121-25, 127
Biospheres: From Earth to Space (Sagan and Margulis), 122-23
Bio-technique/bionics, 15-16, 29, 31, 32, 42, 50, 120, 131
Biotic community, 104
Bird, Eric L., 13
Black, Jeremy, 66
Black Mountain College, 48
Bliss, Arthur, 32
Bomb shelters, 84-85, 118
Bonnemaison, Sarah, 28
Boulding, Kenneth E., 97
Bowie, David, 116
Brand, Stewart, 5-6, 90-91, 94, 102, 113-16, 133
Braungart, Michael, 129
Brave New World (A. Huxley), 27, 34-35
Brazil, 3
Breuer, Marcel, 9, 11, 33, 133
Britain. *See* Great Britain; London Bauhaus
British Empire, 3, 127. *See also* Great Britain
British Interplanetary Society, 88
Brower, David, 115

INDEX > 179

Brown, Jerry, 93
Bryan, John, 148n48
Budberg, Moura, 13
Building Centre (London), 25
Building Research Station (Watford), 28
Burtin, Will, 53, 56

Cabin ecology: and biotechnologies, 93-95; British research on, 118-19; Chermayeff on, 39-40; and human ecology in spaceflight, 84-89; and space program, 6, 7, 39-40, 84-89, 93; and Spaceship Earth, 6, 7, 39-40, 96-101, 106, 108, 127-28. *See also* Closed ecosystems; Space exploration and colonization
Cage, John, 68
California Institute of Technology, 100
Cambridge University, 118-20
Can Our Cities Survive? (Sert), 51
Cape Cod, 109, 110, 111
Capitalism, 24-25, 30, 38, 46, 105, 110, 125. *See also* Industrialism
Carreño, Miro, 56
Carrying capacity, 86, 91, 101-2, 106, 110, 121, 125, 128
Cartography, 54-55, 58-67, 70, *following p. 82*
Caton, Joseph Harris, 143n15
Cave city, 32-36
Chahroudi, Day, 118
Chermayeff, Serge, 10, 13, 39-40, 134, 155n37
Chicago Art Institute, 56
Chicago Institute of Design, 41-42, 48, 56, 70, 126
China, 110
The City of Tomorrow (Le Corbusier), 31
Civil Aeronautics Administration, 153n59
Civil defense program, 84-85
Civil rights movement, 76, 101
Clarke, Arthur C., 88
Closed ecosystems: and Arab oil embargo (1973-74), 102-3; autonomous houses as, 68, 118, 119; bio-shelter as, 118; Biosphere 2 in Arizona as, 113, 116, 121-25, 127; capsule syndrome in ecological architecture, 116-21; Chermayeff and Alexander on, 39-40; and Grumman Corporation, 117; Integral Urban House (Berkeley, Calif.) as, 118; life-support system for, *following p. 82*, 87-88, 118, 121; McHarg's space buoy as, 107; and New Alchemy Institute, *following p. 82*, 108-12, 118, 119, 127, 137; Ouroboros project at University of Minnesota, 128-29; and space cabin ecology, 6, 7, 39-40, 84-89, 93, 95-101, 106, 108, 127-28; space stations as, 89-93; and Spaceship Earth, 96-101; submarines as, 39-40, 84; underground bomb shelters as, 84-85, 118; Yeang's "space craft" as, 120-21, 127
Club of Rome, 78, 91, 92, 101
Co-Evolution Quarterly, 113-16, 133
Coates, Wells, 10, 134
Cold war, 69-72, 84-85, 101, 129
Colonies in Space (Heppenheimer), 92
Colonization of space, 5-6, *following p. 82*, 83-95, 102, 113-16, 163n30. *See also* Space exploration and colonization
Color analysis, 62-63
The Color Harmony Manual (Jacobsen and Gropius), 62-63
Colorado, 55-59, 66, 79, 119, 126
Columbia University, 96
Commerce Department, U.S., 71-72, 79
Commoner, Barry, 111
Community and Privacy (Chermayeff and Alexander), 39-40
Computers, 74-76, 78
Conference on Space Colonization, First, 90-91
Congressional Subcommittee on Space Science and Applications, 91
Consolidated Aircraft Corporation, 52
Container Corporation of America, 41, 54, 55-57, 59, 62, 64-65, 67, *following p. 82*
Contemporary design, 5
Cook, Rick, 130
Cooke, Dennis, 89
Costa Rica, 109

Counterculture movement, 78-82, 90, 101-3, 113-16, 122
Cousteau, Jacques, 114
Crow, Ben, 66
Cyborgs, 93-95

Daly, Herman, 111
Darwin, Charles, 108
Daston, Lorraine, 159n28
Datacity-Metatown project, 34
Dawkins, Richard, 83
DDU. *See* Dymaxion Deployment Unit (DDU)
De Chardin, Pierre Teilhard, 88
De Kooning, William, 56
De Seynes, Philippe, 97-98
Design for the Real World (Papanek), 81
Design with Nature (McHarg), 103, 105-8
Designing with Nature (Yeang), 120-21
DEW Line, 71
Distant Early Warning Line (DEW Line), 71
Domes, 68, 70-76, 81, 110
Doxiadis, Constantinos A., 103, 134
Drop City, Colorado, 79, 119
Dubai, 130
Dubos, René, 93-94
Duncan, Ronald Aver, 31, 32
Dymaxion Deployment Unit (DDU), 48
Dymaxion maps, 68, 70, 77
Dymaxion technologies, 45-48, 73, 79, 82

Earth and Man World Atlas, 66
Earth Day, 67, 97
Earth First!, 101
Earth-works art, 54
Ebeling, Siegfried, 16
Eco-arks. *See* Ark projects
Eco-cells, 121
The Ecological Context (McHale), 98
Ecological design: and Biosphere 2 in Arizona, 113, 116, 121-25, 127; and British Empire, 3; capsule syndrome in ecological architecture, 116-21; current trends in, 128-31; and ecological engineering generally, 5-6; and environmentalism of Gropius, 37-40; and film *Things to Come*, 29-36, 41; Fuller on ecological design revolution, 81; and Fuller's domes, 68, 70-76, 81; and garden city, 11-12, 33; historical overview of, 1-8; impact of space exploration and colonization on, 5-7, 116-25, 127-28; and Integral Urban House (Berkeley, Calif.), 118; and London Zoo, 18-23, 26, 29; and McHarg, 40, 103-8; and Moholy-Nagy, 13-18; and nature writings, 8; and neo-biological civilization, 7; and New Alchemistry Institute, *following p. 82*, 108-12, 118, 119, 127, 137; postcolonial analysis of, 6; and South America, 3. *See also* Bauhaus school; Closed ecosystems; Ecological engineering; Ecology
Ecological engineering: and Biosphere 2 in Arizona, 122-25, 127; and cabin ecology, 6, 7, 39-40, 84-89, 93, 95-101, 106, 108, 127-28; and cyborgs, 93-95; and film *Things to Come*, 29-36, 41; O'Neill and space stations, 89-93; and space research generally, 5-7; and space stations, 89-93. *See also* Ecological design; Ecology
Ecological philosophers, 8
Ecological Society of America, 85
Ecological synergy, 70
Ecology: Arcadian tradition in, 4, 6, 8, 90; cabin ecology, 6, 7, 39-40, 84-89, 93, 95-101, 106, 108, 127-28; definitions of, 46, 47; and energetic systems theory, 85-86; Fuller on, 46-48; holistic ecology, 104-6; managerial approach to, 4, 90, 97, 103, 109, 119-20; meaning of term, 8; Muhlenberg on connections between other disciplines and, 105; of privacy, 39-40; soil ecology, 15; and space colonization, 83-95; and Wells, 13, 27, 30-31, 33-36. *See also* Environmentalism; and headings beginning with Ecological
Economics, 97, 105
Ecumenopolis, 103

INDEX < 181

Ehrlich, Anne, 115
Ehrlich, Paul, 78, 100, 101-2, 115
Einstein, Albert, 92
Energy: Bayer on, 64; conservation of, 93, 94; ecology and energetic systems theory, 85-86; Fuller on, 47-48, 73-74, 79; and Integral Urban House (Berkeley, Calif.), 118; and Leadership in Energy and Environmental Design (LEED), 130; McHarg on, 107-8; and New Alchemy Institute, 110-11; solar cells and solar energy, 102, 115, 117, 118, 119, 121, 128
Energy Basis for Man and Nature (H. T. Odum and E. Odum), 99
England. *See* Great Britain; London
Bauhaus
England and the Octopus (Williams-Ellis), 12
Environment, Power and Society (H. T. Odum), 99, 118
Environmentalism: and Bayer, 60, 64-67; and Fuller, 69, 78-80; and Gropius, 37-40; and Leadership in Energy and Environmental Design (LEED), 130; and McHale, 98; and McHarg, 108; and space exploration and colonization, 83-84, 91-93, 114-16, 122; and Spaceship Earth, 98, 100-101; and Synergia Ranch, New Mexico, 122; vision of, 115. *See also* Ecology
Epcot Center, 2, 93
Ernst, Max, 50
Eton Portrait (Moholy-Nagy), 17
Evolutionary biology, 21-23, 27, 108, 120
The Extended Phenotype (Dawkins), 83

Farming in space. *See* Space farms
Films: by Huxley, 26; by Moholy-Nagy, 17, 18; on space exploration, 88; *Things to Come* and H. G. Wells, 29-36, 41, 148n42, 148n48
Fish, James M., 96
Fish farming, 109, 110
Fisher, Irving, 158n7
Fisher, Ronald A., 23, 27

Fletcher, John C., 91
Food for astronauts, 86-87, 91, 107
Ford, Henry, 45
Foreign Office Architects, 130
Foreman, David, 100-101
Fortune magazine, 70
Foster, Norman, 68, 130
4D Company, 45-46
4D tower house, 45
Francé, Raoul H., 15-17, 134
Frankfurt-Römerstadt housing, 16
Franzen, Ulrich, 97
Frazer, John, 118, 120
Freeman, S. David, 111
Freud, Sigmund, 42, 49
From Eco-Cities to Living Machines: Principles of Ecological Design (Todd and Todd), 111-12
Fry, Edwin Maxwell, 10, 25-26, 38, 134
Fuller, Richard Buckminster: awards and honors for, 73; biographical information on, 42-43, 134; and Biosphere 2 in Arizona, 122; at Chicago Institute of Design, 42, 48; as cold war designer for military-industrial complex, 69-72; and counterculture and environmental movements, 78-82; death of, 82; domes and superdomes of, 68, 70-76, 81, 110; and Dymaxion maps, 68, 70, 77; Dymaxion technologies of, 45-48, 73, 79, 82; on ecological design revolution, 81; and ecological engineering generally, 5; on ecology, 46-48; employment history of, 43-44; on energy, 47-48, 73-74, 79; and 4D Company, 45-46; on GRUNCH (Gross Universe Cash Heist), 82; at Harvard University, 42, 43; influence of, 68-69, 78-82; as journalist for *Life* and *Fortune* magazines, 51-52, 70; lecture tour by, 72-76; marriage of, 43; and Minni-Earth (or Geoscope), 75-77; naval experience of, 43-44, 46, 47, 69, 71, 73, 77, 78-79, 84; on politics, 80-81; and population growth, 44-48, 73, 77-80; as *Shelter* editor, 42; and Simulation Cen-

Fuller, Richard Buckminster *(continued)* ter, Southern Illinois University, 76-78; at Southern Illinois University, 75, 76-78, 81, 98; and space program and space colonies, 94, 95, 114, 127; and Spaceship Earth, 69, 74-76, 78-82, 96-97; teaching positions held by, 42, 48; on Vietnam War, 81; Web sites on, 82, 157n4; and World Game, 68, 76-78, 80, 82; world map by, 70; on world problems, 73-74, 81-82; writings by, 47, 73, 74-75, 81, 97, 98

Functionalism: and Bauhaus school generally, 3-5; and Bayer, 50, 51; of cave design, 32-36; and Lubetkin, 19; and Moholy-Nagy, 14-16, 19, 50, 51; and relationship between science and architecture, 7; and Sullivan's motto "Form follows function," 2, 51

Fundamentals of Ecology (E. P. Odum), 89

Furniture, 11

The Future of the Future (McHale), 94-95

Gaia atlases, 66
Gaia hypothesis, 83, 100, 114, 122-23, 125
Galison, Peter, 7
Gardens, 11-12, 33, 105
Gardiner, Henry, 59
Geddes, Norman Bell, 32
Geddes, Patrick, 40, 104
Geddes, Robert, 104
Gehry, Frank, 130
Gemini program, 74, 119. *See also* Space exploration and colonization
Genetical Theory of Natural Selection (Fisher), 23
Geodesics, Inc., 71
Geography, 54-55, 58-66, 70, *following p. 82*
Geoscope, 75-77
German Micrological Society, 15
Giedion, Sigfried, 56
Giurgola, Romaldo, 104
Glaser, Peter E., 92
Gleaves, Admiral, 44
Gloag, John, 10

Global perspective, 48-53, 69-70, 97-99, 103
Goethe, Johann Wolfgang von, 41, 49, 59, 62
Golley, Frank B., 66, 93
Grass Mound (Bayer), 54, 58
Great Britain: economic depression in, during early 1930s, 24-25, 27; Histon village college in, 25-26; London Bauhaus in, 4, 9-23, 30, 126; London slums in, 22, 24-25, 29; London Zoo in, 13, 18-23, 26, 29; National Plan for, by Nicholson, 25-26; research on cabin ecological systems in, 118-19. *See also* specific architects and designers

Green Building Council, U.S., 130
"Green" buildings, 119-20
Gropius, Walter: and Bauhaus school, 3, 9-10, 14, 17, 36, 37, 49, 58, 97, 131; biographical information on, 134; and biological sciences, 3; and *Color Harmony Manual*, 62; as environmentalist, 37-40; farewell dinner for, in London, 9, 10, 12, 13, 23, 142n1; and film *Things to Come*, 33; and garden city, 12; at Harvard School of Design, 9, 37-40, 126; and Histon village college, 25-26; and Huxley, 3, 9; and Isokon Building, 10; and MARS, 10; and Michael Reese Hospital, University of Chicago, 39; and PEP (Political and Economic Planning Organisation), 25; retirement of, from Harvard, 104; in United States, 9, 36-40, 126; writings by, 14

Grumman Corporation, 117-18
Grumman Lunar Module, 117
GRUNCH (Gross Universe Cash Heist), 82
Guide to the New World (Wells), 36

Haldane, John B. S., 23
Haraway, Donna, 95
Hardin, Garrett, 79-80, 100, 115
Hardwicke, Cedric, 35, 149n56
Harriman, Richard L., 100, 101-2
Harvard School of Design, 9, 37-40, 62, 103-4, 126

Harvard Society for Contemporary Art, 45
Harvard University, 42, 43, 122
Hawes, Phil, 121, 122
Hazelrigg, George, 90
Health. *See* Public health
Heidegger, Martin, 8
Henderson, Hazel, 115
Henderson, Lawrence J., 40
Henson, Keith and Carolyn, 91
Heppenheimer, Tom A., 92-93, 164n31
Hewlett, Anne, 43
The High Frontier (O'Neill), 91-92
Hill, David, 56
Hippies. *See* Counterculture movement
Histon village college, 25-26
Holford, William G., 13
Holism and Evolution (Smuts), 104
Holistic ecology, 104-6
Household prototype technologies, 117
Howard, Ebenezer, 33
Howlett, J. Monroe, 44
Human ecology. *See* Ecology
Huntington, Ellsworth, 64
Hurt, Philip D'Arcy, 21
Huxley, Aldous, 27, 34-35
Huxley, Julian: biographical information on, 134-35; and ecological design generally, 127; and Gropius, 3, 9, 131; *If I Were Dictator* by, 26-29; and London Zoological Society, 13, 26; nature film by, 26; and PEP (Political and Economic Planning Organisation), 26, 27; and Studio of Animal Art, 26; and TVA, 28-29; writings by, 23, 26-30, 147n25

If I Were Dictator (J. Huxley), 26-29
Industrialism: and Bayer, 54; Chermayeff on, 39; in film *Things to Come*, 32; Fuller and inventive industrialization, 46, 47; Fuller and military-industrial complex, 69-72; Gropius on, 37, 49; McHarg on, 105-6; Nicholson on, 25; O'Neill and military-industrial complex, 89; Pike on, 118-19; space colonization as alternative to, 6, 92, 106. *See also* Capitalism

Information theory, 87
Innes, Donald Q., 163n30
Institute of Ecotechnics, 122
Integral Urban House (Berkeley, Calif.), 118
International Style architecture, 5, 33
Isobar Club, 11
Isokon Building, 10-13, 142n3
Isokon Laminated Furniture series, 11

Jacobsen, Egbert, 55, 57, 62
Jasanoff, Sheila, 99
Jet Propulsion Laboratory, 100

Kahn, Louis I., 104, 105
Kandinsky, Wassily, 49-50
Kennedy, John F., 84, 85
Kesey, Ken, 115
King, Ynestra, 100-101
Kirk, Andrew G., 94
Korda, Alexander, 13, 26, 31, 33, 35, 135
Korda, Michael, 35
Korda, Vincent, 33
Kubrick, Stanley, 88
Kurlansky, Mark, 101

Landscape architecture, 103-8, 127
Lavin, Sylvia, 7
Lawn Road Flats, 10-13, 142n3
Le Corbusier, 19, 31, 32, 33, 34
Le Sourd, Homer, 69
Leadership in Energy and Environmental Design (LEED), 130
LEED (Leadership in Energy and Environmental Design), 130
Léger, Fernand, 33, 56
Lemco, Blanche, 104
Leopold, Aldo, 8
Life magazine, 51-52, 70
The Life of the Lobster, 18
Life-support system, *following p. 82*, 87-88, 118, 121
Lifeboat ethics, 100
Light modulator, 16
Light-Play: Black-White-Gray (Moholy-Nagy), 17

The Limits to Growth (Club of Rome), 78, 91, 92, 101, 115
Lissitzky, El, 50
Lockheed Missiles and Space Company, 99, 117-18
London Bauhaus, 4, 9-23, 30
London slums, 22, 24-25, 29
London Zoo, 13, 18-23, 26, 29
London Zoological Society, 13, 20, 21, 23, 29
Lovejoy, Thomas, 122, 124
Lovelock, James E., 83, 100, 116, 135
Lubetkin, Berthold, 10, 13, 18, 19, 21, 26, 135
Lunar space program. *See* Moon

Maas, Winy, 34, 130
Macy, Christine, 7, 28
Malaysia, 120
Malthus, Thomas, 44-45, 60, 78, 79-80, 92, 122
Managerial approach to ecology, 4, 90, 97, 103, 109, 119-20
Manes, Christopher, 100-101
Margalef, Ramón, 89, 116, 123, 135
Margulis, Lynn, 100, 111, 114-16, 122-23, 135
Markoff, John, 78
Mars, 83, 88, 93, 100, 102, 113, 123, 124-25. *See also* Space exploration and colonization
MARS (Modern Architectural Research Group), 10-12, 20
Marx, Roberto Brule, 3
Marxism, 81
Masdar City, Abu Dhabi, 130
Massachusetts Institute of Technology, 118
Massey, Raymond, 35
Matless, David, 12
May, Ernst, 16
Mayer, Hannes, 4
McCormick, Jack, 105
McDonagh, James E. R., 13
McDonough, William, 129
McHale, John, 66, 94, 98, 122, 135

McHarg, Ian, 5, 40, 103-8, 120, 127, 135
McLarney, William, 109, 111
McQuaid, Kim, 88
Meadows, Dennis, 115
Meerson, Martin, 104
Mendel, Gregor, 23
Menzies, William C., 148n48
Mercury program, 74. *See also* Space exploration and colonization
Michael Reese Hospital, University of Cicago, 39
Mies van der Rohe, Ludwig, 4
Military-industrial complex, 69-72, 89. *See also* Industrialism
Mills, Stephanie, 115
Minni-Earth (or Geoscope), 75-77
Mitchell, Peter Chalmers, 21-23, 135
Modern Architectural Research Group (MARS), 10-12, 20
Modernist design, 5, 7, 12, 22-23
Moholy-Nagy, László: and Bauhaus school, 2, 9, 10, 13-18, 36, 37, 40-42, 58, 126; biographical information on, 136; and biological sciences, 2, 13-18, 29, 41-42, 50, 70, 126, 131; at Chicago Institute of Design, 41-42, 56, 70, 126; death of, 41; definition of architecture by, 16; and film *Things to Come*, 32-33, 41; and functionalism, 14-16, 19, 50, 51; and Isokon Building, 10, 142n3; and learning from nature's workshop, 13-18; and MARS, 10; and New Bauhaus, 40-41, 56; and organic as important, 143n15; patrons of, 41, 56; photographic art and film projects of, 17-18, 61; on Sullivan's motto "Form follows function," 2, 51; in United States, 36, 37, 40-42, 126; writings and publications by, 14-15, 17, 33, 41
Moholy-Nagy, Sibyl, 32, 63, 65
Moon, 85, 87-91, 93, 102, 106, 115, 117. *See also* Space exploration and colonization
Moore, Henry, 56
Morgenstern, Oscar, 77
Morgenthaler, Fritz, 105
Morpho-ecologic design, 130

Muhlenberg, Nicholas, 105
Muir, John, 8
Mumford, Lewis, 12, 16, 115
Murphy, Michelle, 119-20
Museum of Modern Art, 18, 37, 50-53, 60, 153*n*59
Myers, Jack, 84, 86

Naess, Arne, 8
Nakagawa, Masato, 59
NASA: and Apollo program, 99, 117, 119; and cabin ecology, 87-89; and Gemini program, 74, 119; and ground control in Houston, 76, 108; and Grumman Corporation, 117; and Lovelock's detection devices for planetary exploration, 100; and Mercury program, 74; and Princeton University conferences, 85; and solar energy, 102; and space colonies, *following p. 82*, 89, 91, 116; and Spacelab program, 90. *See also* Space exploration and colonization
"A National Plan for Great Britain" (Nicholson), 25-26
National Science Foundation, U.S., 102
Naval Academy, U.S., 43-44, 77
Navy, U.S., 43-44, 46, 47, 69, 71, 73, 77-79, 84, 100
Nelson, Mark, 122
Neuman, John von, 77
Neutra, Richard, 7, 129
New Alchemy Institute, *following p. 82*, 108-12, 118, 127, 137
The New Architecture of the London Zoo, 18
New Bauhaus, 40-41, 56
New Bodleian Library, Oxford, 27-28
New Mexico, 121, 122
New State of the World atlases, 66
The New Vision (Moholy-Nagy), 14-15
New York, 130
New York Botanical Garden, 124
Nicholson, Edward Max, 13, 25-27, 136
Nine Chains of the Moon (Fuller), 47
Noah's Ark, 92, 102, 109-10, 112, 113, 123. *See also* Ark projects

NORAD (North American Aerospace Defense Council), 76, 95

Ocean Arks, *following p. 82*, 112, 137
Ocean Group, 130
Odum, Elisabeth, 99
Odum, Eugene P., 85-87, 89, 92, 99, 121, 124, 125, 136
Odum, Howard T., 85-86, 99, 118, 123, 125, 136
Office of Civil Defense (OCD), 84-85
Office of Naval Research, 85
Oil embargo (1973-74), 102-3
Olgyay, Aladar and Victor, 11
O'Neill, Gerard Kitchen, 89-93, 114-16, 122, 136
Operating Manual for Spaceship Earth (Fuller), 74-75, 97, 98
Ostwald, Friedrich Wilhelm, 62
Ouroboros project (University of Minnesota), 128-29
Overpopulation. *See* Population growth
Overy, Paul, 12
Oxford University, 17, 27-28
An Oxford University Chest (Moholy-Nagy), 17

Paepcke, Elisabeth, 41, 66
Paepcke, Walter, 41, 55-57, 59, 64-66, 136, 155*n*37
Pang, Alex Soojung-Kim, 71
Papanek, Victor, 81
Park, Katharine, 159*n*28
Passuth, Kristina, 14
Patten, Bernard C., 87
Pearce, Peter, 70
PEP (Political and Economic Planning Organisation), 13, 25-27
Perkins, G. Holms, 104
Phillips, John Frederick Vicars, 104-6, 136
Photography: of Ansel Adams, 57; of Earth from space, 99, 103, 106; by Moholy-Nagy, 17, 61
Phylogenesis, 130
Picon, Antoine, 7

Pike, Alexander, 118-19
Pirages, Dennis C., 102
Political and Economic Planning Organisation (PEP), 13, 25-27
Pollution. *See* Environmentalism
Pont Foundation, 90
Ponte, Alessandra, 7
The Population Bomb (Ehrlich), 78, 101
Population growth: and AIDS, 101; Bayer on, 60, 61; and carrying capacity, 86, 91, 101-2, 106, 110, 121, 125, 128; Club of Rome's *Limits to Growth* on, 78, 91, 92, 101, 115; Ehrlich on, 78, 100, 101-2, 115; Fuller on, 44-48, 73, 77-80; and Hardin's lifeboat ethics, 100; Malthus on, 44-45, 60, 78, 79-80, 92, 122; space exploration and colonization as response to, 91, 92, 123; Yeang on, 120
Porter, Ros, 66
Postcolonial analysis, 6
Prince Edward Island, Canada, 109, 110-11
Princeton University, 85, 87, 89-92
Pritchard, Jack, 11
The Private Life of the Gannets, 26
Progressive Era, 45-46
Proxmire, William, 91
Public health, 12, 21, 28
Pusey, Fredrick, 148n48

Rand McNally, 66
Ray, Man, 56
Recycling, 56, 67, *following p. 82*, 84, 93, 94, 122
Reichstag (Berlin), 68
Reilly, Charles Herbert, 13
Renner, George, 58
Rivera, Diego, 45
Roberts, Walter Orr, 123
Rome, Adam, 38-39
Rosenzweig, Martin, 59
Rousseau, Jean-Jacques, 8, 95
Royal Institute of Architecture, 31
Rudolph, Paul, 97, 104

Safdie, Moshe, 129
Sagan, Carl, 114

Sagan, Dorion, 122-23
Samuel, Godfrey, 10, 20-21, 137
Samuel, Herbert, 20
San Diego State University, 109
Satellite technology, 91-92, 114
Schaffer, Simon, 158-59n28
Schulten, Susan, 70
Schumacher, Ernst F., 115
Schumpeter, Joseph, 45
Schwartz, Julius, 165n2
Science. *See* Biological sciences; Ecological engineering; Ecology; Space exploration and colonization
Science fiction and futurist writings, 27, 34-35, 88
The Science of Life (Wells, Huxley, and Wells), 30, 147n25
Scott, James C., 69
Self-contained design. *See* Closed ecosystems
Seminar on Environmental Arts and Sciences, 66
Serres, Michel, 95
Sert, José Luis, 51
Shand, Morton, 10
The Shape of Things to Come (Wells), 31
Shelter, 33, 41-42
Shelter Design and Analysis (Office of Civil Defense), 84-85
Sick-building syndrome, 120
Simberloff, Daniel S., 92
Simulation Center, Southern Illinois University, 76-78
"Sky-Roads" traveling exhibition, 153n59
Slums, 22, 24-25, 29, 51
Small Is Beautiful (Schumacher), 115
Smith, Eric Alden, 115
Smithsonian Institution, 124
Smuts, Jan Christian, 104
Snyder, Gary, 115
The Social Life of Monkeys (Zuckerman), 21
Sociobiology (Wilson), 83
Soft Technology Research Community, 119
Soil ecology, 15
Solar cells and solar energy, 102, 115, 117, 118, 119, 121, 128

Solar Energy Laboratory, 118
Soleri, Paolo, 102, 116, 137
South Africa, 104, 105, 106
South America, 3
Southern Illinois University, 75, 76-78, 81, 98
Space Biosphere Ventures Inc., 122
Space buoy, 107
Space cities, 114
Space Colonies (Brand), 114
Space exploration and colonization: Apollo program, 99, 117, 119; and Biosphere 2 in Arizona, 113, 116, 121-25, 127; and biotechnologies, 86-87, 93-95, 127, 128; and cabin ecology, 6, 7, 39-40, 84-89, 93, 95-101, 106, 108, 127-28; and carrying capacity, 86, 91, 106, 121, 128; colonial terminology used in space program, 5-6; and counterculture movement, 101-3, 113-16; and cyborgs, 93-95; debate on space colonies among co-evolutionists, 113-16; and environmentalism, 83-84, 91-93, 114-16, 122; and food for astronauts, 86-87, 91, 107; Gemini program, 74, 119; human ecology in spaceflight, 84-89; impact of, on ecological design, 5-7, 116-25, 127-28; life support system for lunar base, *following p. 82*, 87-88; lunar space program, *following p. 82*, 85, 87-91, 93, 102, 106, 115, 117; of Mars, 83, 88, 93, 100, 102, 113, 123, 124-25; and McHarg's space buoy, 107; Mercury program, 74; O'Neill and space stations, 89-93; opposition to space colonies, 100, 115-16; and space colonies, 5-7, *following p. 82*, 83-95, 102, 113-16, 163*n*30; Spacelab program, 90. *See also* NASA
Space farms, 91, 92-93, 107, 109-10
Space modulator, 16, 18
"Space Oddity" (Bowie), 116
Space stations, 89-93
Spacelab program, 90
Spaceship Earth: and Biosphere 2 in Arizona, 113, 116, 121-25, 127; and Boulding, 97; and cabin ecology, 96-101, 106, 108, 127-28; and carrying capacity, 101-2, 106, 110, 125, 128; and counterculture movement, 101-3; at Disney's Epcot Center, 2, 93; and ecological ethic, 93-94, 96-101, 127-28; first published mention of, 165*n*2; and Fuller, 69, 74-76, 78-82, 96-97; and Gaia hypothesis, 100; and McHale, 98; and McHarg, 103-8; and New Alchemy Institute, *following p. 82*, 108-12, 118, 127, 137; and photos of Earth taken from moon, 99, 103, 106; and space colonies, 6, 163*n*30; and United Nations, 97-98; and Ward, 97
Spaceship Earth (Ward), 97
Stanford University, 91
State University of New York at Buffalo, 94
Stein, Richard, 111
Steiner, Hadas, 7, 22-23
Stephan, Nancy Leys, 3
Stevenson, Adlai, 65, 97
Stout, Randall, 130
The Street Markets of London (Moholy-Nagy), 17
Structure in Nature Is a Strategy for Design (Pearce), 70
Studio of Animal Art, 26
Submarines, 39-40, 84, 114
Suburban sprawl, 38-39, 120
Sullivan, Louis, 2, 51
Summerson, John, 12, 14
Superdomes, 72. *See also* Domes
Survival of the fittest, 108, 120-21
Synergetics (Fuller), 81
Synergia Ranch, New Mexico, 122, 124

Taylor, Frederick, 45
Tecton Company, 19-21, 23, 26
Tennessee Valley Authority (TVA), 28-29, 103
Testament (Wright), 3
Things to Come, 29-36, 41, 148*n*42, 148*n*48
Third World Atlas (Crow and Thomas), 66
Thomas, Alan, 66
Thompson, Emily, 7
Thompson, William Irwin, 116
Thoreau, Henry David, 8, 37
Thorne, Oakleigh, 66

Thorne Ecological Foundation, 66
Todd, John, 109-11, 116, 127, 137
Todd, Nancy, 111-12, 116
Towards a New Architecture (Le Corbusier), 34
Turner, Fred, 90
Turtle Island (Snyder), 115
TVA, 28-29, 103
Twain, Mark, 86
2001: A Space Odyssey, 88

U Thant, 97
Underground bomb shelters, 84-85, 118
United Nations, 53, 75-76, 97-98, 118
United Nations Conference on the Environment (1972), 118
University Corporation of Atmospheric Research, 123
University of Arizona, 91, 121, 124
University of Chicago, 39, 46
University of Georgia, 86
University of Ghana, 104
University of Minnesota, 128-29
University of Pennsylvania, 103-4
University of Southern California, 67
University of Texas, 84
Urban environmental renewal, 37-40
Urban sprawl, 38-39, 120
Urbanism (Le Corbusier), 31

Vajk, J. Peter, 164n31
Vale, Brenda, 119
Vale, Robert, 119
Van Eyck, Aldo, 105
Vanderbilt, Tom, 85
Városmajor Utca housing project (Budapest), 11
Venturi, Robert, 104
Vevers, Geoffrey, 21
Vietnam War, 81, 101
Vision in Motion (Moholy-Nagy), 41

Waddington, Conrad H., 13
Wald, George, 115

Wank, Roland, 29, 137
War games, 76-78
Ward, Barbara, 97
Warshall, Peter, 115
Waters, Mike, 95
Week-End Review, 25-27
Wellesley-Miller, Sean, 118
Wells, Frank, 33
Wells, G. P., 147n25
Wells, H. G.: and Bauhaus school, 12-13; biographical information on, 137; and ecology, 13, 27, 30-31, 33-36; and film *Things to Come*, 29-36, 41, 148n42, 148n48; Harwicke on, 149n56; and London Zoo, 29; and PEP, 26; writings by, 30-31, 36, 76, 147n25, 148n41
Wells, Malcolm, 129
Wheaton, William L. C., 104
Wheeler, Monroe, 51-52
White, Gilbert, 8
Whitney Museum of American Art, 82
Whole Earth Catalog, 79, 90, 94, 102, 113, 133
Wichita House, 48
Williams-Ellis, Clough, 12, 13, 137
Wilson, Edward O., 83, 92, 122, 137
Wood, Denis, 157-58n7
The Work, Wealth and Happiness of Mankind (Wells), 30-31
World Brain (Wells), 76
World Game, 68, 76-78, 80, 82
World Geo-Graphic Atlas (Bayer), 54-55, 58-66, following p. 82
World Resources Inventory, 98
World War I, 43-44, 46, 49, 69
World War II, 48, 51-53, 69-70
Worster, Donald, 66
Wright, Frank Lloyd, 3, 46
Wymore, Wayne A., 91

Yeang, Kenneth, 120-21, 127

Zoos, 13, 18-23
Zuckerman, Solly, 20-21, 137